Lecture Notes in Computer Science 4295

Commenced Publication in 1973
Founding and Former Series Editors:
Gerhard Goos, Juris Hartmanis, and Jan van Leeuwen

James D. Carswell   Taro Tezuka (Eds.)

Web and Wireless Geographical Information Systems

6th International Symposium, W2GIS 2006
Hong Kong, China, December 4-5, 2006
Proceedings

 Springer

Volume Editors

James D. Carswell
Digital Media Centre
Dublin Institute of Technology
Dublin 2, Ireland
E-mail: jcarswell@dit.ie

Taro Tezuka
Graduate School of Informatics
Kyoto University
Yoshida-Honmachi, Sakyo
Kyoto, 606-8501, Japan
E-mail: tezuka@dl.kuis.kyoyo-u.ac.jp

Library of Congress Control Number: 2006936624

CR Subject Classification (1998): H.2, H.3, H.4, H.5, C.2

LNCS Sublibrary: SL 3 – Information Systems and Application, incl. Internet/Web and HCI

ISSN	0302-9743
ISBN-10	3-540-49466-9 Springer Berlin Heidelberg New York
ISBN-13	978-3-540-49466-9 Springer Berlin Heidelberg New York

Springer is a part of Springer Science+Business Media

springer.com

© Springer-Verlag Berlin Heidelberg 2006
Printed in Germany

Typesetting: Camera-ready by author, data conversion by Scientific Publishing Services, Chennai, India
Printed on acid-free paper   SPIN: 11935148   06/3142   5 4 3 2 1 0

Preface

These proceedings contain the papers selected for presentation at the sixth edition of the International Symposium on Web & Wireless Geographical Information Systems held in Hong Kong during December 2006. This symposium was intended to provide an up-to-date review of advances in both theoretical and technical development of Web and Wireless Geographical Information Systems (W^2GIS). It was the sixth in a series of successful events beginning with Kyoto 2001, and alternating locations annually between East Asia and Europe. It now represents an ever-increasing spatially aware geotechnology research community.

Now in its sixth year, W2GIS has matured in name from a "workshop" to a full 2-day "symposium" – recognition by the field as a forum for quality dissemination and discussion on the latest research and development achievements in the domain. The number of papers received for this symposium demonstrates not only the growing importance of this field for researchers but also the growing impact these developments have in the daily lives of all citizens.

From well over 130 submissions, 72 papers were initially selected as being directly in scope with the symposium, and from these, 24 papers (33%) were selected for final presentation and inclusion in the proceedings. Each paper received three reviews and was ranked accordingly. The accepted papers cover a wide range of topics from the Semantic Web, Web personalization, contextual representation and mapping to querying in mobile environments, mobile networks and recent developments in location-based services and applications.

We had the privilege of having a distinguished invited talk by Yufei Tao, Department of Computer Science of the City University of Hong Kong. The best paper from the symposium was selected by the Steering Committee and invited for an extended journal publication by *Transactions in GIS*.

We wish to thank the authors that contributed to this workshop for the high quality of their papers and presentations and the support of Springer LNCS. We would also like to thank the Program Committee for the quality and timeliness of their evaluations. Finally, many thanks to the Steering Committee for providing continuous advice.

October 2006

James D. Carswell
Taro Tezuka

W2GIS 2006 Symposium Committee

Symposium Chairs

J.D. Carswell, Dublin Institute of Technology, Ireland
T. Tezuka, Kyoto University, Japan

Industrial Chair

K.J. Li, Pusan National University, Korea

Local Chair

H. Lin, The Chinese University of Hong Kong, SAR China

Steering Committee

M. Bertolotto, University College Dublin, Ireland
C. Claramunt, Naval Academy, France
B. Huang, The Chinese University of Hong Kong, SAR China
C. Vangenot, EPFL, Switzerland

Program Committee

P. Agouris, University of Maine, USA
M. Arikawa, University of Tokyo, Japan
A. Bouju, University of La Rochelle, France
T. Cheng, University College London, UK
K. Clarke, University of California, Santa Barbara, USA
N. Cullot, University of Burgundy, France
M.L. Damiani, DICO - University of Milan, Italy
R.A. de By, ITC, Netherlands
M. Duckham, University of Melbourne, Australia
M. Egenhofer, NCGIA, USA
M. Gahegan, Penn State, USA
Y. Ishikawa, Nagoya University, Japan
C.S. Jensen, Aalborg University, Denmark
M.A. Kang, University Clermont-Ferrand, France
H.A. Karimi, University of Pittsburgh, USA
B. Köbben, ITC, Netherlands

Y.J. Kwon, Hankuk Aviation University, Korea
R. Laurini, INSA Lyon, France
H. Li, Chinese Academy of Sciences, China
S. Li, Ryerson University, Canada
M. Mainguenaud, INSA Rouen, France
P. Muro-Medrano, Universidad de Zaragoza, Spain
S. Nittel, University of Maine, USA
B.C. Ooi, National University of Singapore, Singapore
D. Papadias, HKUST, Hong Kong, SAR China
G. Percivall, Open Geospatial Consortium, USA
D. Pfoser, Computer Technology Institute, Greece
M. Raubal, University of Munster, Germany
A. Rodriguez, University of Concepcion, Chile
M. Schneider, University of Florida, USA
S. Spaccapietra, EPFL, Switzerland
K. Tanaka, Kyoto University, Japan
Y. Tao, City University of Hong Kong, SAR China
G. Taylor, University of Glamorgan, UK
Y. Theodoridis, University of Piraeus, Greece
A. Voisard, Fraunhofer ISST and FU Berlin, Germany
R. Weibel, University of Zurich, Switzerland
S. Winter, The University of Melbourne, Australia
O. Wolfson, University of Illinois at Chicago, USA
S. Yi, Huazhong University of Science and Technology, China
I. Zaslavsky, San Diego Super Computer Center, USA
A. Zipf, Mainz University of Applied Sciences, Germany

External Reviewers

Keith Gardiner, Dublin Institute of Technology, Ireland
Elias Frentzos, University of Piraeus, Greece
Kostas Gratsias, University of Piraeus, Greece
P. Álvarez, University of Zaragoza, Spain
J. Nogueras, University of Zaragoza, Spain
J. Lacasta, University of Zaragoza, Spain
S. Martínez, University of Zaragoza, Spain

Sponsors

K.C.Wong Education Foundation, Hong Kong
PointI.com, Korea
Research Support Unit, Dublin Institute of Technology, Ireland

Table of Contents

Session 3 - Wayfinding, Mobile and Wireless GIS

Session 4 - W2GIS Personalization and Agent

Session 5 - Data Management and Data Retrieval Methods

Session 6 - Semantic Geo-spatial Web and Ubiquitous W2GIS

Putting Location-Based Services on the Map

Michael Grossniklaus, Moira C. Norrie, Beat Signer, and Nadir Weibel

Institute for Information Systems
ETH Zurich
8092 Zurich, Switzerland
{grossniklaus, norrie, signer, weibel}@inf.ethz.ch

Abstract. Location-based services for users on the move provide a convenient means of filtering information based on current geographical position. However users also often want to retrieve or capture information associated with past or future locations. We show how new technologies for interactive paper can be used to augment conventional paper maps with location-based services using a combination of user tracking and pointing to the map to specify location.

1   Introduction

Location-awareness is an important aspect of context-awareness and can thus be exploited to reduce the information bandwidth in mobile systems by delivering only information relevant to the current user situation. This is important due to not only limitations of mobile devices, but also the fact that users may be involved in parallel activities or need to react quickly to a given situation. However, users in mobile environments often want to retrieve or capture information relevant to a particular location either before or after the visit, possibly as part of a planning or decision-making process. For example, tourists may use time in a cafe to enquire about restaurants near the next location that they plan to visit or to write entries in their travel journal.

Tasks that involve the planning and reviewing of routes and visits within a city require an overview of the spatial layout of a city rather than just localised maps, together with the ability to easily view areas other than the current location. A well-known disadvantage of current mobile devices is the fact that the small screen size makes it awkward to perform such tasks. While some projects have used Tablet PCs to increase available screen size, this clearly restricts mobility. Conventional paper maps, on the other hand, satisfy the requirements of mobility as well as providing the required spatial overview. In addition, they have advantages in terms of readability in outdoor environments, especially by users collaborating, and can be easily annotated.

We show how digital pen and paper solutions can be used to provide a range of location-based services based on interactions with a paper map and audio feedback delivered via a text-to-speech engine. Our system offers flexible positioning modes as locations can be specified either *implicitly* based on current user

J.D. Carswell and T. Tezuka (Eds.): W2GIS 2006, LNCS 4295, pp. 1–11, 2006.

location obtained from a position sensor, *explicitly* by having the user point to the map with the pen or by using coordinates *stored* in an application database.

In previous work, we have developed a general framework that enables digital pen and paper solutions to be used not only for the digital capture of handwriting, but also for real-time interaction with a vast range of digital services. In this paper, we show how the resulting concept of interactive paper can be used to augment a paper map with four well-known types of location-based services that provide information about places of interest, events and also help the user find locations on the map. As an additional location-based service, handwritten information can be captured and associated with either the user's current location or a location pointed to on the map. For each of these services we will examine in detail the kinds of positioning modes that they support by describing the functioning of selected tasks a user can accomplish with our system.

In Sect. 2, we motivate the use of interactive paper maps as an interface to location based services in mobile environments and discuss related work. Section 3 presents the underlying technologies for interactive paper and Sect. 4 then describes the location-based services accessible through the map. Details of the system implementation are presented in Sect. 5 and concluding remarks are given in Sect. 6.

2   Motivation and Related Work

Tourists make extensive use of maps before, during and after city visits. Tasks performed during a visit have been categorised as *locator*, *proximity*, *navigation* and *event* tasks [1]. Locator tasks are concerned with questions such as *"Where am I?"* and *"Where is X?"*. Proximity tasks deal with finding objects or people located near to the user's current location. Navigation tasks involve route finding either from the user's current location to a given location or object, or between any pair of specified objects/locations independent of the current location. Event tasks deal with finding information about what happens at a given place.

Early digital maps tended to focus on navigation, but recent ethnographic studies on the use of maps by tourists emphasise the need to support a much wider range of tasks [2]. Large maps provide a tourist with an overview of the features of a city and the spatial relationships that help tourists locate themselves, identify potential areas of interest and keep track of places visited. Various forms of annotation are often used to highlight locations and routes and note names of facilities such as restaurants.

Tourism is a domain with considerable potential for the use of mobile technologies and a number of projects have developed PDA-based tourist guides, with Cyberguide [3] and GUIDE [4] among the earlier examples. While some projects have focussed on navigation, others have addressed issues such as context-awareness, interaction, visualisation or collaboration. Studies show that tourists spend a lot of time comparing and combining information, but it is difficult to support these activities on small digital mobile devices. Also interaction is often awkward as it is tedious to input large amounts of information. For this reason, some projects have opted for the use of Tablet PCs [4, 5], enabling them to provide much richer information displays and services to the user. Clearly the problem with these devices based on current technologies is that they are relatively large and heavy. Another approach is to investigate means for accessing supplementary and dynamic information as well as advanced digital services in a more natural way. The use of pen and paper as an interface can be seen as a *natural interaction* and various projects have investigated the use of paper maps either for accessing digital information [6, 7] or for the capturing of information [8]. Digital pen and paper functionality offered by the Swedish company Anoto [9] provides an easy-to-use solution for the digital capture of handwriting on paper and several projects have used these to build annotation systems, such as for example in ButterflyNet [10]. These systems are much more convenient for users than inputting annotations digitally, even when using handwriting on digital devices.

For all four categories of map-based tasks defined at the beginning of this section, location plays a key role and hence location-awareness based on user tracking is an important feature of many mobile tourist information systems. Some systems allow tourists to specify an arbitrary location on the map [11], or to select a location from a list of the last recorded positions [4, 12]. However, in many cases, the specification of the location is only meant to be used as a fallback solution if no other positioning system is available and is not supported as a real feature of the system. Other systems such as Rasa [13] or NISMap [8] support explicit positioning on a map, but they do not provide implicit tracking of positions, since they are not meant to be used in a mobile environment. Nevertheless, pointing to map positions is a very convenient way of specifying, not only where one wishes to go in seeking directions, but also past and future locations. For example, in planning activities for the evening ahead, a tourist may want navigation information from their hotel to a restaurant or information about events close to the hotel without them actually being located at the hotel. It is therefore important that user location can be specified either explicitly, implicitly or through coordinates stored in an application database. If implicit positions from a location sensor are used, it has to be possible to use several alternative tracking technologies that are combined seamlessly.

Another desirable feature of mobile tourist information systems is the possibility to capture information. The Graffiti system [14], for instance, allows tourists on the move to create electronic notes and associate them with a location. In [15], mobile phones equipped with RFID receivers were used to track user positions and publish location-based pictures, videos or annotations.

Our goal was to develop a complete solution that could support all four categories of map-based tourist tasks, exploiting the advantages of the combination of implicit, explicit and database-driven position tracking, as well as providing means for tourists to record their travel experiences. In the next sections, we describe how this can be achieved through the implementation of interactive paper maps based on a general cross-media content publishing system that we have developed.

3   Interactive Paper

A number of digital pens are now commercially available based on Anoto functionality. These technologies are able to track the position of a pen on paper through a combination of a special dot pattern printed on paper that encodes position information and a camera inside the pen as illustrated in Fig. 1(a).

The dot pattern, which is almost invisible, encodes (x,y) positions in a vast virtual document space. Camera images are recorded and processed in real-time giving up to 100 (x,y) positions per second. The technology was originally developed for the digital capture of handwriting and several pages of handwriting can be captured and stored within the pen before being transferred to a PC via a Bluetooth or USB connection. Hitachi Maxell and Logitech have recently released pens based on Anoto functionality that can also be used in streaming mode where position information is transmitted continuously. This enables the pens to be used for real-time interaction as well as writing capture.

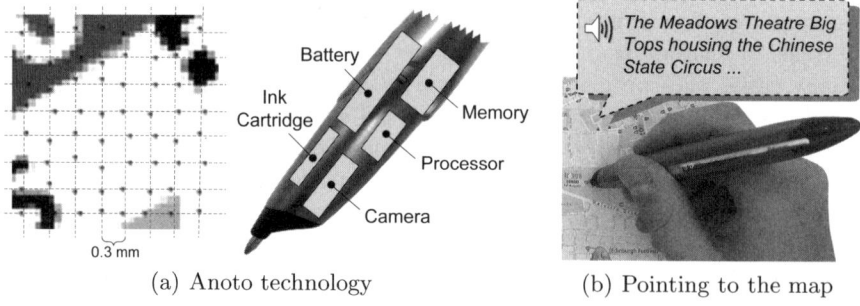

(a) Anoto technology   (b) Pointing to the map

Fig. 1. Interactive paper maps

We have developed a general link service for interactive paper [16, 17], which enables active areas to be defined on paper and bound to digital resources such as images, videos or web pages, or to services such as a text-to-speech engine, a database system or an application such as Microsoft PowerPoint. Thus, the digital pen on paper can be used in much the same way as a mouse would be used during web browsing to activate links to static documents as well as trigger specific application calls.

Using these technologies, we have developed interactive paper maps to support tourists on the move based on a text-to-speech output channel. Interaction with the map provides access to a range of location-aware services and users can also annotate the map or link annotations written in a separate document to positions on the map. Maps have been developed both for general tourist information in the city of Zurich and, specifically, to support visitors to the Edinburgh Festivals based on a map showing festival venues as presented in Fig. 1(b). We use the latter example in the rest of this paper to describe the functionality, operation and implementation of the system in detail.

4   Location-Based Services

In this section, we detail the various types of location-based services and inter-actions that can be supported by interactive maps. Figure 2 shows part of a map developed for visitors to a festival, where services can be accessed by touching the pictograms at the top. In addition, certain services may be accessed by di-rectly pointing to a location on the map. For example, in the case of the festival system, pointing to a location anywhere on the map accesses information about the venue closest to that location.

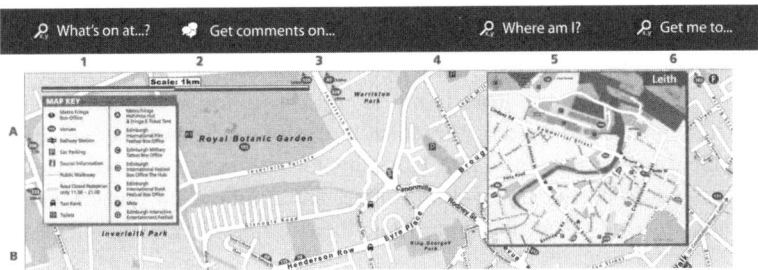

Fig. 2. Accessing location-based services on the map

Table 1 shows the services offered by our system grouped under the four categories described earlier. In addition to these services we added annotation functionality that will be described later in this section. A main contribution of our system is its support to acquire positional context information in different ways and the seamless transition between these input modalities. A first method to provide input for a location-based service involves real-time positional infor-mation provided by a sensor such as a Global Position System (GPS) device. This *implicit* tracking of positional information and its application for access-ing location-based services represents the most common input form in existing location-aware applications. However, in our system a user can also provide lo-cation information *explicitly* by pointing with the digital pen to a specific loca-tion on the interactive map. Finally, positions that are *stored* in the application database can be used. Note that our architecture not only supports these three positioning modes for providing location information, but based on the actual service, the user can also freely combine them. To give an example, let us have a look a the navigation service *Get me (from X) to Y* shown in Tab. 1. If a user does not explicitly provide information about X, the current GPS position is taken as input parameter. However, the user can also specify X by pointing to the map with the digital pen or it can be derived from information currently accessed from the application database. The same mechanism can be applied to define the target Y for this specific navigation service. We will now examine in more detail the positioning modes used in the services our system offers.

Locator tasks enable tourists to find the current position of people or objects on the map. To find their own position, location is specified implicitly by a

Table 1. Positioning modes for location-based services

Service		Explicit	Implicit	Stored
Locator	*Where am I?*		✓	
	Where is X?			✓
Proximity	*What is near X?*	✓	✓	✓
	Who is near X?	✓	✓	✓
Navigation	*Get me (from X) to Y.*	✓	✓	✓
Event	*What is going on at X?*	✓	✓	✓
Annotations		✓	✓	✓

tracking sensor. On the other hand the position of another person or the location of a sight could be stored in the database and retrieved from it. For example, in the festival system, the coordinates of all festival venues are stored in the application database. When the map locator service is activated to find the position where an event is taking place, the application queries the festival database and retrieves the position of the location associated to this event.

Proximity is defined as the task of locating objects or people nearby. The use of three distinct positioning modes allows us to extend this definition to locate entities either in proximity to the user, a location pointed to on the map or an object stored in the application database. For this task, our system supports all three kinds of location modes. As an example, after touching the pictogram of the *What's on at...?* service, a user can explicitly specify a position on the map with the pen and then the system retrieves all events nearby based on the venue coordinates stored in the database.

Most mobile navigation systems provide functionality to guide tourists from their current position provided by a sensor to an explicitly specified location or attraction. Moreover they normally adapt the directions if tourists do not follow the indications. Our system supports the *Navigation* task, by allowing tourists to get directions based on coordinates from the database as well as implicit and explicit positioning modes. To plan future activities, a tourist may specify both the starting point and destination on the map. On the other hand, if tourists want directions to a location, the current position will be used as the starting point. Navigation may even be useful in the other direction to provide a route from a specific point on the map to the tourist's current position in order to give directions to someone else. In the festival system that service is based on landmark navigation since this was considered to be appropriate for tourists in an unknown city [18]. Of course, it is also possible to integrate other navigation services into our framework.

Accessing information about what is going on at a particular location is defined as an *Event* task. Information about events may be accessed in very different ways. One tourist may be interested in what is going on right now or what is on at a given theatre in the evening. Our location-based services may be used to find events happening close to the current location or a location specified on the map and to answer questions like *"What is nearby?"* or *"What is going on at the location X?"*. Such services were central to our festival system and our interactive map provided an easy and natural means of accessing them and could be used on its own or together with other interactive documents such as the event brochure.

As described earlier, a novel feature of our system is the use of audio feedback to guide users when working with pen and paper. As an example of such an interaction based on paper based pen input and audio output, let us look at the locator process. The locator algorithm first checks whether the position of the search location is on the map. If this is the case, the locator computes the quadrant on the map which contains the location based on the physical size of the map and the coordinate system used to express the position of the location. It is important to note that our system supports maps of any size and in the case of the festival system we offered both A3 and A4 maps as well as a mini-map printed on a paper bookmark. A series of iterative steps is then used to direct the user's pen to the correct position on the map. Whenever the user touches the map with the pen, the locator computes the distance D between the current position of the pen's nib and the desired destination. It then indicates to the user how to get the pen closer to the desired location by telling them the corresponding offset in terms of how many centimetres they have to move up, down, left or right. Once the distance D between the pen's position and the position of the desired location falls below a specific threshold P, the system switches over to the nearness mode where the exact distance is no longer told to the user, but they are guided using the approximate *"You are close! Move a little bit to the right"*. Once the tourist has successfully moved their pen to the search location on the map, the locator process informs the user that they have found it and then terminates.

Since our system is basically composed of a map and a pen, annotations are a simple way of integrating comments and notes. In order to highlight interesting locations or activities, tourists may write comments on the map, annotating directly the area of the city or the specific sight of interest. Tourists may insert notes or comments and link them with a particular area or location on the map. The comment added is stored in XML for later retrieval, either by the same tourist or by another tourist interested in the same location. As shown in Tab. 1, linking handwritten notes to a position on the map is in fact another location-based service offered by our system. The annotations may be either automatically linked to the current tourist's position provided by the GPS sensor or explicitly bound to a specific location on the map. Once annotations have been inserted into the system, the interactive paper map also allows tourists to later retrieve the written comments by pointing to the *Get comments on...* pictogram. If the user also specifies a position on the map, the system retrieves comments bound to locations within a certain range from it. Otherwise the system gets the current position as specified by the GPS sensor, retrieving comments linked to the locations nearby.

The services and interactions presented so far have primarily focused on information that is delivered by the system to the user. However, by offering the functionality to store annotations made on paper in digital form, our system supports a different kind of interaction that provides the user with the possibility to input and store information from the map directly in the system.

5   Implementation

To support the provision of location-based services, our system uses three distinct information sources to track positions of the users and objects—a GPS sensor, positions coming from the map and coordinates stored in the application database. The GPS sensor defines positions by means of latitude, longitude, elevation and time, whereas the positions on the map are defined as a set of (x,y) coordinates. The use of maps of different sizes and formats does not allow the definition of a clear and unique system, but rather encodes positions in many different scales. Moreover, the position of locations, sights, buildings and people could be stored in an database using different application-specific coordinate systems.

The architecture of our system, shown in Fig. 3, takes the different coordinate systems into consideration and enables the delivery of location-based information using a seamless switching between the implicit and explicit positions, as received by the sensors, or application-specific positions from the *Application Database*.

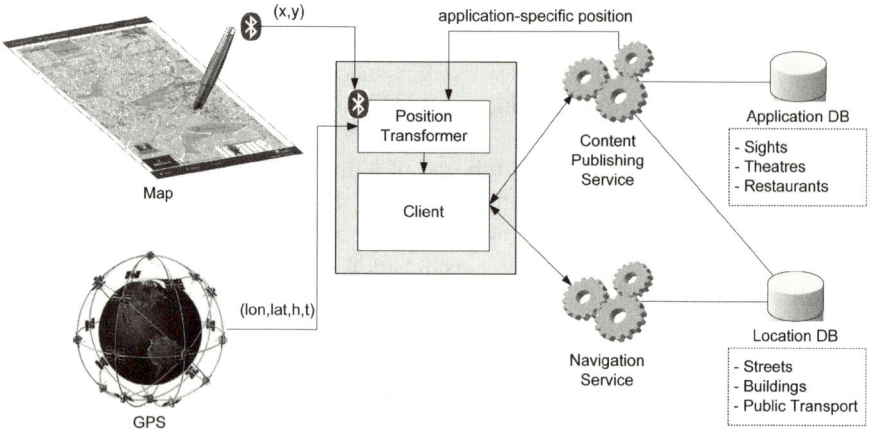

Fig. 3. General architecture of the interactive map system

The *Client* is connected to a GPS sensor, to a Bluetooth digital pen and to the application database through a *Position Transformer* component. In order to use the implicit current position of the user, explicit positions whenever the pen touches the map or the application-specific positions coming from the database, the position transformer acts as a proxy and transforms the positions into a unique coordinate system, which may vary depending on the application. For the festival application the standard Universal Transverse Mercator (UTM) system was used, as it is currently a good approximation of spatial coordinates on a two-dimensional plane. The GPS position is retrieved when needed by the different location-based services. When receiving an (x,y) position from the map, depending on the requested functionality, the Client may contact either the *Content Publishing Service* or the *Navigation Service*. The Content Publishing Service is responsible for retrieving information from the Application Database. Depending on the user's preferences, it then publishes the location-based information by means of special templates which define how to present the output to the user. In our festival system, the Content Publishing Service is responsible for delivering VoiceXML output, which is then interpreted by a text-to-speech engine present on the client device. In other applications, we used visual output channels such as PDAs, wall-mounted displays and also head-mounted displays.

If the Navigation Service is invoked, the system generates routing information based on the positions specified by the tourist (implicit, explicit or stored) and on the landmarks stored in the *Location Database*. The information returned is then processed by the Content Publishing Service which generates the appropriate output format.

The control flow of the application is depicted in Fig. 4. The interaction is started by the pen which touches the map in order to select a location-based service. If a position on the map is directly selected and no specific service is active at the moment, the default service will be selected, which returns information about the nearest location. Otherwise, if the user first selects a specific service (e.g. *What's on at...?*), the context will be set to this service and a second interaction with the map is needed in order to specify a location. Alternatively, if the user does not select any explicit location, the implicit current position will be retrieved from the GPS sensor. After transforming the positions received by the sensors, the system selects the right service and forwards the positions to it. If needed the system will then contact the application database in order to retrieve stored positions. Finally the system informs the user about the results of the issued request.

To handle the capture of handwritten notes, capture areas may be defined anywhere within the document. In the case of the map, it is possible to define a capture area that covers the entire map. If a user starts to write within a capture area, the system is able to detect from the prolonged contact of the pen and paper that it is a writing rather than a pointing gesture and switches automatically to capture mode. A general capture component collects all positional information coming from the pen and generates an XML representation of the captured data which can be processed by an Intelligent Character Recognition (ICR) engine to get its textual representation or transformed to a JPEG image. In the case of annotating the map to link a comment with a specific location, the bounding box of the captured strokes is computed and the physical location represented by the centre of the bounding box is associated with the comment. However, if a user is writing a comment in a capture area that is not located on the map and does not select an explicit position on the map within a given amount of time, the user's current position provided by the GPS sensor is automatically associated with the captured note.

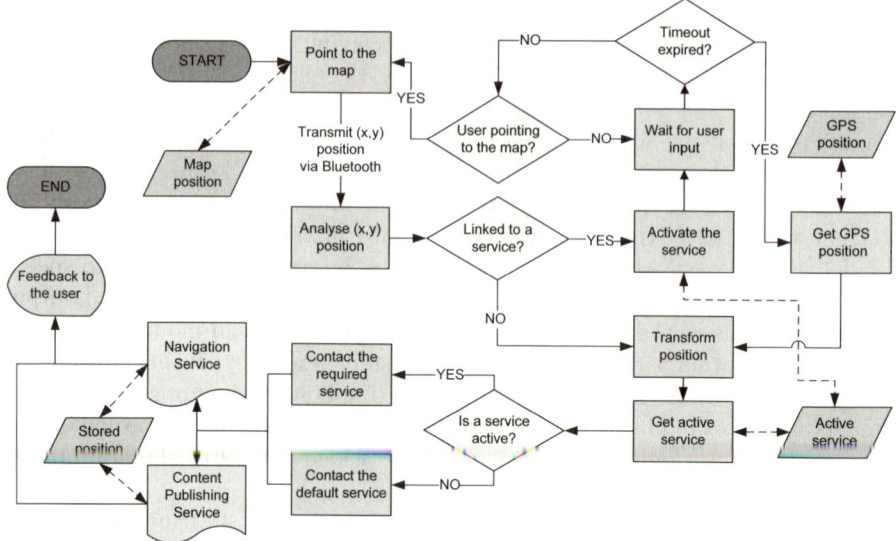

Fig. 4. Application control flow

6   Conclusions and Future Work

We have shown how technologies for interactive paper can be used to provide
a range of location-based services in mobile environments based on interaction
with printed maps together with user positioning systems. Our system supports
three positioning modes—explicit, implicit and stored—that can be combined
depending on the type of location-based service. To interact with users, we com-
pensate for the static nature of the printed map by using audio feedback to guide
the usage of pen on paper.

Although, in this paper, we have focussed on the use of interactive maps in
mobile environments, clearly they may also be of use in other settings, especially
to support forms of collaboration and flexible annotation that are difficult to
mimic in digital worlds.

References

1. Reichenbacher, T.: Adaptive Concepts for a Mobile Cartography. Journal of Geographical Sciences (2001) 43–53
2. Brown, B., Laurier, E.: Designing Electronic Maps: An Ethnographic Approach. In
Meng, L., Zipf, A., Reichenbacher, T., eds.: Map Design for Mobile Applications.
Springer Verlag (2004)
3. Abowd, G., Atkeson, C., Hong, J., Long, S., Kooper, R., Pinkerton, M.: Cyberguide: A Mobile Context-Aware Tour Guide. Wireless Networks **3** (1997) 421–433
4. Cheverst, K., Davies, N., Mitchell, K., Friday, A., Efstratiou, C.: Developing a
Context-Aware Electronic Tourist Guide: Some Issues and Experiences. In: Proceedings of CHI 2000, ACM Conference on Human Factors in Computing Systems,
The Hague, The Netherlands (2000) 17–24

5. Brown, B., Chalmers, M., Bell, M., MacColl, I., Hall, M., Rudman, P.: Sharing the Square: Collaborative Visiting in the City Streets. In: Proceedings of CHI 2005, ACM Conference on Human Factors in Computing Systems, Portland, USA (2005)
6. Fitzmaurice, G.W.: Situated Information Spaces and Spatially-aware Palmtop Computers. Commun. ACM **36**(7) (1993) 39–49
7. Grasso, A., Karsenty, A., Susani, M.: Augmenting Paper to Enhance Community Information Sharing. In: Proceedings of DARE 2000, Conference on Designing Augmented Reality Environments, Elsinore, Denmark (2000) 51–62
8. Cohen, P.R., McGee, D.R.: Tangible Multimodal Interfaces for Safety-Critical Applications. Commun. ACM **47**(1) (2004) 41–46
9. Anoto: Pen and Paper Solutions. (http://www.anoto.com)
10. Yeh, R.B., Liao, C., Klemmer, S.R., Guimbretière, F., Lee, B., Kakaradov, B., Stamberger, J., Paepcke, A.: ButterflyNet: A Mobile Capture and Access System for Field Biology Research. In: Proceedings of CHI 2006, ACM Conference on Human Factors in Computing Systems, Montréal, Canada (2006)
11. Baus, J., Krueger, A., Wahlster, W.: A Resource-Adaptive Mobile Navigation System. In: Proceedings of IUI 2002, 7th International Conference on Intelligent User interfaces, San Francisco, USA (2002) 15–22
12. Pospischil, G., Umlauft, M., Michlmayr, E.: Designing LoL@, a Mobile Tourist Guide for UMTS. In: Proceedings of MobileHCI 2002, 4th International Symposium on Human Computer Interaction with Mobile Devices and Services, Pisa, Italy (2002) 140–154
13. McGee, D.R., Cohen, P.R., Wu, L.: Something from Nothing: Augmenting a Paper-Based Work Practice via Multimodal Interaction. In: Proceedings of DARE 2000, International Conference on Designing Augmented Reality Environments, Elsinore, Denmark (2000) 71–80
14. Burrell, J., Gay, G.K.: Collectively defining Context in a Mobile, Networked Computing Environment. In: Proceedings of CHI 2001, ACM Conference on Human Factors in Computing Systems, Seattle, USA (2001) 231–232
15. Cheng, Y.M., Yu, W., Chou, T.C.: Life is Sharable: Blogging Life Experience with RFID Embedded Mobile Phones. In: Proceedings of MobileHCI 2005, 7th International Symposium on Human Computer Interaction with Mobile Devices and Services, Salzburg, Austria (2005) 295–298
16. Norrie, M.C., Signer, B.: Information Server for Highly-Connected Cross-Media Publishing. Information Systems **30**(7) (2005) 526–542
17. Signer, B.: Fundamental Concepts for Interactive Paper and Cross-Media Information Spaces. PhD thesis, ETH Zurich (2005)
18. Raubal, M., Winter, S.: Enriching Wayfinding Instructions with Local Landmarks. In: Proceedings of GIScience 2002, 2nd International Conference on Geographic Information Science, Boulder, USA (2002) 243–259

Beyond Location Based – The Spatially Aware Mobile Phone

Rainer Simon, Peter Fröhlich, and Hermann Anegg

Telecommunications Research Center Vienna (ftw.),
Donau-City-Str. 1, A-1220 Vienna, Austria
{simon, froehlich, hanegg}@ftw.at

Abstract. An increasing number of mobile phones feature embedded sensors such as GPS receivers, digital compasses or accelerometer-based tilt sensors. In this paper, we present an application framework for building spatially aware mobile applications – applications that visualize, process or exchange geo-spatial information – on mobile phones equipped with such sensors. The core component of the framework is a novel, platform-independent XML data exchange format that describes the geographic vicinity of the mobile device. The format enables a variety of new mobile interaction styles and user interface types – from traditional text-based local search and information interfaces to innovative real-time user interfaces like geo-pointers and smart compasses.

Keywords: Spatial information appliances, personal spatial assistants, location based services, geo-spatial information.

1 Introduction

Almost all information that exists is related to a certain place, area, or location: historical data may be related to buildings or streets; a timetable is related to a bus stop or train station; most content available on the World Wide Web is related to a place or region. In essence, information is almost always – at least to a certain extent – geographical. Web-based mapping applications like Google Maps, Yahoo Maps or Microsoft Windows Live Local, as well as geo browsing applications like Google Earth or NASA World Wind have recently raised the public awareness for the usefulness and the educational, as well as the entertainment value of geographical information.

Even though many mobile applications today offer comparable functionality – such as location-based search and mapping – few have succeeded in conveying a user experience nearly as rich and dynamic as their desktop counterparts. Interestingly, novel and compelling concepts for interacting with geo-spatial information on mobile devices have already been suggested years ago. Based on the technology of Geographic Information Systems, Egenhofer [1] predicted *Spatial Information Appliances* – portable tools for professional users and a public audience alike, relying on fundamentally different interaction metaphors: *Smart Compasses* that point users into the direction of certain points of interest, *Smart Horizons* that allow users to look

J.D. Carswell and T. Tezuka (Eds.): W2GIS 2006, LNCS 4295, pp. 12–21, 2006.

beyond their real-world field of view or *Geo-Wands* – intelligent geographic pointers that allow users to identify geographic objects by pointing towards them.

Our work is motivated by the vision that mobile phones will soon serve as generic hard- and software platforms for a variety of spatial information appliances. The device features necessary to realize this vision are already becoming available: Embedded GPS receivers, digital compasses and accelerometer-based tilt sensors are found in an increasing number of handsets and can be expected to be even more widespread in the near future. What is still missing today is a common application framework: a toolkit that leverages mobile geo-spatial applications by enabling developers to experiment with new interaction metaphors and to prototype user interfaces that offer experiences beyond what is offered by today's applications.

In this paper, we present such a framework, which we developed based on requirements derived from a series of Wizard-of-Oz user tests [3]. The paper is structured as follows: Section 2 discusses related work and explains our concept of spatial awareness. Section 3 lists the technical requirements we derived from the user tests. Section 4 describes our proposed data exchange format that enables rapid prototyping and development of spatially aware mobile applications. Section 5 presents our framework implementation. Section 6 concludes with an outlook on future work.

2   The Spatially Aware Device

Several research projects have investigated mobile interaction with geo-spatial information: GeoNotes [2], Nexus [7], Urban Tapestries [5] or Riot [10], for example, are based on the common idea of attaching digital information to real-world places like a virtual post-it note or graffiti. A handheld device equipped with a location sensor (e.g. a GPS receiver) allows users to consume this location-based information or to actively participate as provider of information. Other projects have experimented with additional sensors beyond GPS: Wasinger et al, for example, presented a PDA-based application that uses GPS and a digital compass to realize Geo-Wand-like two-dimensional pointing functionality [13]. Similar ideas were applied by Mitchell et al [6] in the context of a mobile multiplayer game and Strachan et al [12] in the context of a spatial-audio-based navigation application.

The common idea of creating a digital information space, interconnecting it with the real world through geographical references and using handheld devices as bridges between the real and the virtual space is consistent with our understanding of the spatial information appliance. However, we argue that the concept described by Egenhofer goes further: A spatial information appliance is not just a collection of sensors that measure basic geographical properties, which subsequently serve as search parameters in a spatial database query. Rather, we see spatial information appliances as smart, spatially aware personal geographical assistants: devices which themselves possess locally stored knowledge about the environment around them – its structure, its geometry and its visual appearance – and their own relative position and situation therein. This explicit knowledge enables them to support users in navigation and orientation tasks in real-time and to offer intuitive and compelling user experiences beyond what is offered by today's location-based services.

3 Requirements

Embedded sensors are obviously one essential precondition for enabling spatially-aware applications. A second fundamental prerequisite is a data exchange format that captures and encodes geographical knowledge about the environment. In order to support more usable interfaces to geo-spatial information around the user, the format must model the environment in accordance with the user's perception of space. It must also take into account the special characteristics of mobile devices, i.e. it must be suitable for processing on computationally limited hardware, and it must support a wide variety of different user interface types and interaction styles, since mobile devices differ widely in terms of form factor, input- and output characteristics and feature set (e.g. no sensors vs. embedded GPS and compass).

In a user study, we have tested and compared mock-up user interfaces for different types of spatial information appliances [3]. The results of the study revealed the following three important findings:

- While the traditional map representation still ranked among the most popular forms of user interfaces, ego-centric representations, such as the Geo-Wand or simplistic textual lists of nearby points of interest, were also rated as highly usable and intuitive.
- Users appreciated an explicit indication of the visibility of nearby features and points of interest.
- Orientation-awareness in three dimensions, rather than only in two, was appreciated. This is in accordance with findings presented e.g. by Rakkolainen and Vainio [9].

From the results of the user tests, and from our own experiences during the system specification phase of the project, we derived the following five technical requirements for a geographical data exchange format for spatial information appliances: *generality*, *user-centeredness*, *cross-platform scalability*, *simplicity* and *compactness*.

Generality. The data format should be application-agnostic. Geographic XML formats such as the Geography Markup Language (GML [8]) are based on a Cartesian map metaphor. As our user tests have shown, intuitive spatially aware mobile applications can also be based on distinctively different interaction metaphors. While it is probably not entirely possible to define a data format without a particular user interface scenario in mind, the format should at least allow for a variety of possible solutions beyond maps.

User-centeredness. The metaphors used by the format should conform to the way users perceive their environment. As mentioned, three-dimensional representation has been identified as relevant in this context [9]. A closely related concept that has been found to be of importance in our user studies, and which is lacking an explicit representation in traditional maps or vector geometry formats is *visibility*: We therefore argue that knowledge of the visibility of geographic features from a certain point of view is crucial for enabling spatially aware applications. Computing the visibility from a 3D geometry model is a computationally highly intensive task and difficult to realize on mobile devices with low processing power. The format should therefore contain explicit information about the visibility of features or at least pre-process the data so that visibility computation is simplified.

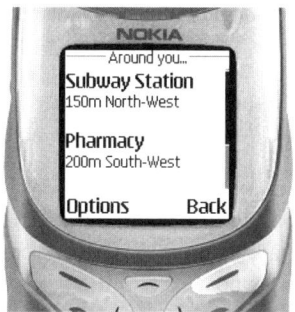

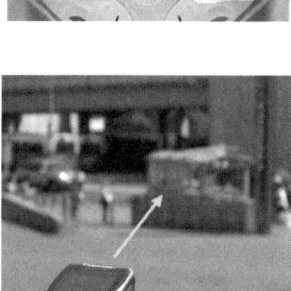

Fig. 1. Possible LVis GUI concepts

Cross-platform scalability. On the one hand, the data format should be explicit: Client devices with low computational power and limited graphical output capabilities should be able produce immediate output without the need for complex arithmetic calculations, e.g. by transforming it to a meaningful text. On the other hand, the format should be rich and detailed: High-end devices should be able to create a more compelling, richer user experience at the expense of higher processing overhead. In particular, the format should carry true structural information about the environment, so that the client can react to data from embedded sensors in real-time, without the need to re-query the application server. For example, the client might be able to re-compute distances, headings or the visibility of certain points of interest locally, as the user moves through the environment.

In order to illustrate the range of scenarios the data format might be required to support, Figure 1 shows concept illustrations of four user interface designs we have tested in our user study. Each design assumes different device capabilities and sensor equipment: In the most basic case (shown top left) only the location of the device is known and no particular computational power is available. In this case, only basic textual output is presented. The upper right image shows a basic *Smart Compass* interface that indicates nearby points of interest by arrows on an ego-centric compass rose. This or a similar scenario might be realized on a device with built-in compass and GPS. The lower left image denotes the principle of the *Geo-Wand*, which might be realized on a device equipped with differential GPS and compass [11]. Finally, the lower right image shows a concept for a more sophisticated user interface: Using a combination of differential GPS, compass and a 2-axis tilt sensor, a simplified augmented reality (AR) user interface might be envisioned, where labels are superimposed over the phone's live camera image to indicate points of interest on the screen.

Simplicity. The concepts and metaphors used by the format to model the environment should be simple and easy to understand. This is predominantly a technical requirement, since it reduces the entry barrier for application developers previously not involved with geographic applications, or who have little expertise in geographical or graphical vector formats like SVG.

Compactness. Last but not least, the format should be as compact as possible, so that airtime use and response time are minimized.

In the following section, we present the *Local Visibility Model – LVis* in short (pronounced "Elvis") – our proposed XML data format for spatially aware mobile applications, which satisfies the five requirements listed above.

4   Local Visibility Model

The *Local Visibility Model* represents a simplified, ego-centric geometric model of the local environment around a geographic position. Unlike a typical map, the LVis describes the structure of the environment in three dimensions, rather than only in two. Unlike other geographic encoding formats, it also contains implicit information about the visibility of geographic features. Since the LVis is XML-based and relies on standard units and measurements (i.e. meters and decimal degrees) meaningful textual output can be produced by simply styling the XML accordingly. Due to the fact that the LVis uses a polar coordinate notation, user interfaces that would normally require transformations from Cartesian to polar coordinate space (such as Smart Compasses or Geo-Wands) can be realized with considerably reduced development effort.

4.1   LVis XML Syntax

The LVis distinguishes two types of spatial entities: *content* and *geometry*. The *content* of the LVis is formed by a collection of geographic markers. Each marker indicates an arbitrary point of interest (poi) and describes it with application-specific meta-data. Compared to a map, the function of a marker corresponds to that of an icon drawn on the map, e.g. to indicate a certain landmark. Since the LVis uses polar coordinates, each marker location is defined by its distance (in meters) and its heading and elevation angle (in decimal degrees) relative to the LVis center position.

Only markers that are visible from the LVis center position are contained in the LVis. Figure 2 shows an example of the syntax used to express a marker in the LVis XML format.

```
<poi id="poi6" description="Snack Bar" d="117" hdg="102" elev="5"/>
```

Fig. 2. LVis poi ("point of interest") XML element

The LVis *geometry* model describes the geometry of geographic features in the area around the LVis center. Our design goal was to keep modeling concepts and data structures as simple as possible. In particular, it should be possible to derive meaningful and user-understandable textual output by simply applying a transformation to the XML code (e.g. using XSLT). This approach restricted us from relying on traditional vector geometry formats, where geometry is typically described using polygons or geometric primitives in Cartesian 2D or 3D space. These formats require complex processing to produce visual output and are not human-readable as such.

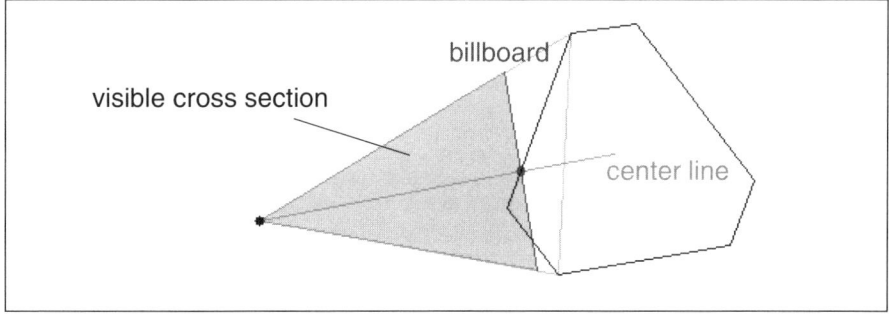

Fig. 3. LVis billboard approximation (top-down view)

In order to satisfy the conflicting requirements of describing the surrounding environment geometry sufficiently detailed and yet in a simple format that can be interpreted with an absolute minimum of processing, we decided to model the environment based on a *billboard* metaphor: Each geographic feature, such as a building, is approximated by a flat, rectangular wall, facing towards the LVis center. In essence, the LVis geometry model can be thought of as a 360-degrees panoramic "cardboard cutout" version of the vicinity, much like a movie set where the environment is not made up of solid buildings, but instead of building facades. Each billboard is defined by the distance, the heading and elevation angle of its center point relative to the LVis center, its width (in decimal degrees) and height, and application-specific descriptive meta-data. A textual description of the environment (e.g. "There is an office building to the North-East in 250 meters") can therefore be produced without complex arithmetic computations or coordinate transformations.

Figure 3 illustrates the principle of how an LVis billboard is computed: The illustration shows a top-down view of an arbitrary building. First, the visible cross section of the building, seen from the LVis center, is determined. The billboard is computed by intersecting the cross section's center line with the building shape and computing a normal to the center line.

```
<billboard   id="Bldg09"   description="Tech   Gate   Tower"   d="42"
  hdg="125" width="41" elev="28"/>
```

Fig. 4. LVis billboard XML element example

An example of the XML syntax used to describe an LVis billboard is shown in Figure 4. A complete LVis code sample that encodes 3 points of interest and 4 billboards around a specific geographic location is shown in Figure 5.

```
<lvis long="16.413598392" lat="48.232866">
    <poi   id="poi3"   description="Telecommunications   Research   Center
    Vienna office" hdg="203" d="15" height="16"/>
    <poi    id="poi4"    description="    Pharmacy"    hdg="93"    d="29"
    height="4"/>
    <poi    id="poi7"    description="Bus    stop"    hdg="91"    d="152"
    height="2"/>

    <billboard id="Bldg09" description="Tech Gate" hdg="223" d="13"
    width="154" height="26"/>
    <billboard   id="Bldg04"   description="Subway   station"   hdg="94"
    r="152" width="13" height="4"/>
    <billboard id="Bldg01" description="Tech Gate Tower" hdg="125"
    d="42" width="41" height="21"/>
</lvis>
```

Fig. 5. Complete LVis code sample

5 Framework Implementation

Our current application framework consists of a Java library that implements the functionality needed to compute the visible content markers (points of interest) and the billboards for fully and partially visible buildings. The framework relies on a 2.5D environment block model, i.e. each building in the model must be represented by a 2D building footprint shape and a single height parameter per building. The model used with our test setup consists of sample data covering an area of roughly 1x1 km around our office premises (an architecturally rather unusual business district in the North of Vienna, Austria). The data is stored in a PostgreSQL/PostGIS database.

After querying the database, the framework uses a basic scan-line algorithm to compute a 360-degrees line of sight from the LVis center position (see Gardiner and Carswell [4] for a comparable approach). The integration of an advanced visibility detection algorithm using *guided visibility sampling* [14] is currently under development.

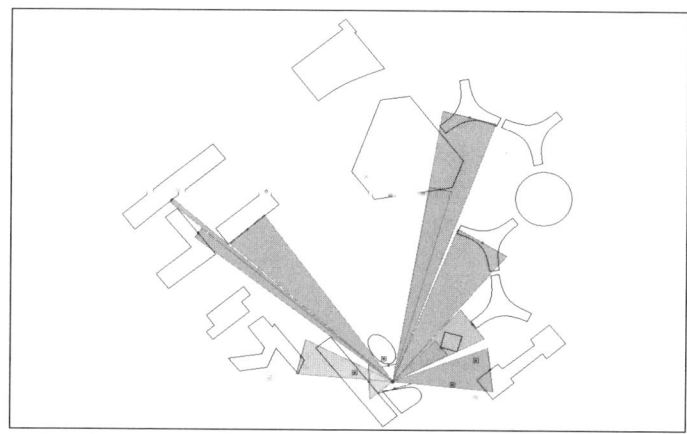

Fig. 6. LVis computation result (top down view)

Fig. 7. LVis computation (panoramic view)

Figures 6 and 7 show the result of an example LVis computation performed by the implementation: Figure 6 shows the computed LVis in a top-down view; Figure 7 shows a 360 degrees panoramic visualization of the output. The visible content markers are shown as small dark rectangular dots. To make the billboards easier to recognize in the top-down view, a "viewing beam" extends from the LVis center to each billboard, with beams of lighter color indicating higher billboards. In the panoramic view, rectangles of lighter color represent more distant billboards.

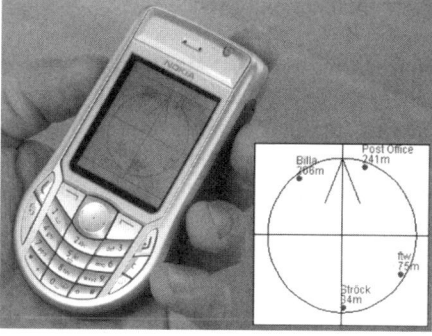

Fig. 8. Outdoor function trials (Notebook and phone)

First function trials were carried out with the framework running locally on a Notebook, connected to a Bluetooth GPS receiver, as shown in the left image in Figure 8. Further trials were carried out with a handheld test device: Figure 8, right, shows a mobile phone (connected to the same Bluetooth GPS) with an example *Smart Compass* interface. For this test, the framework was integrated with an open source web server, allowing the mobile device to communicate with the framework using over-the-air HTTP requests.

6 Future Work

The focus of our current and upcoming activities lies on application prototyping and user testing with functional applications. Since mobile devices with a full set of sensors are not yet readily available on the market, we are working on a custom sensor hardware module for our experiments. The module will combine differential GPS, a compass and a 2-axis tilt sensor in a self-contained Bluetooth unit.

The application scenarios to be implemented and tested include those depicted in Figure 1 and a scenario where users actively participate in generating location-based information, i.e. where the LVis is locally manipulated and re-submitted to the application server. While we have so far investigated the theoretical performance and accuracy that can be achieved by our system [11], the tests will reveal the practical limits that occur under real world conditions.

Also, the tests will further clarify whether the modeling concepts used in the LVis are effective in supporting more usable mobile interfaces to geo-spatial information. Insights in how people use our applications will influence and guide the further evolution of the LVis XML format. Modeling of hidden features and an assessment of the billboard metaphor's suitability for non-urban environment are among the topics that will be addressed:

Hidden features. In its current version, the LVis does not contain information about geographical features that are not directly visible from the user's current position. As has been found in our user trials, it is desirable to have this information available in many application scenarios. We expect the user tests to deliver valuable insights into how fully and partially hidden features can be efficiently modeled in accordance with users' perception of space.

Non-urban terrain. While the billboard metaphor has so far proven useful for modeling urban environment, it is unclear whether the same metaphor can also be applied to geographic features in rural terrain. Tests will show whether it is reasonable to model topographical features like mountains or hills using billboards, or whether an alternative model should be applied.

Acknowledgments

The work presented in this paper is part of the Telecommunication Research Center Vienna's project *P2D – Point to Discover* (http://p2d.ftw.at/). *P2D* is funded by mobilkom Austria, Siemens Austria and the **K***plus* competence centre program.

References

1. Egenhofer, M. J.: Spatial Information Appliances: A Next Generation of Geographic Information Systems. 1st Brazilian Workshop on GeoInformatics, Campinas, Brazil, 1999.
2. Espinoza, F., Persson, P., Sandin, A., Nyström, H., Cacciatore. E., Bylund, M.: GeoNotes: Social and Navigational Aspects of Location-Based Information Systems. Proceedings of Ubicomp 2001, Atlanta, Georgia, September 30 - October 2, 2001.
3. Fröhlich, P., Simon, R., Baillie L., Anegg, H.: Comparing Conceptual Designs for Mobile Access to Geo-Spatial Information. Proceedings of Mobile HCI 2006, Espoo, Finland, September 12-15, 2006.
4. Gardiner, K., Carswell, J. D. Viewer-Based Directional Querying for Mobile Applications, Third International Workshop on Web and Wireless Geographical Information Systems (W2GIS2003), IEEE CS Press, Rome, Italy, 2003.
5. Lane., G.: Urban tapestries: Wireless networking, public authoring and social knowledge. Personal Ubiquitous Computing. July 2003. Vol. 7, no. 3-4, pp 169–175.
6. Mitchell, K., McCaffery, D., Metaxas, G., Finney, J., Schmid, S., Scott, A.: Six in the City: Introducing Real Tournament – A Mobile Context-Aware Multiplayer Game. Proceedings of the 2nd Workshop on Network and System Support for Games, ACM Press, 2003.
7. Nicklas, D., Mitschang, B.: The NEXUS Augmented World Model: An Extensible Approach for Mobile, Spatially-Aware Applications. Proc. of the 7th Int. Conf. on Object-Oriented Information Systems, Calgary, 2001.
8. OpenGIS Geography Markup Language (GML) Encoding Specification. http://portal.opengeospatial.org/files/?artifact_id=4700
9. Rakkolainen, I., Vainio, T.: A 3D City Info for Mobile Users, Computers & Graphics, Vol. 25, No. 4, Aug. 2001, Elsevier.
10. Reid, J., Hull, R., Cater, K., Clayton, B. Riot! 1831: The design of a location based audio drama. Proceedings of UK-UbiNet 2004, October 2004.
11. Simon, R., Kunczier, H., Anegg, H.: Towards Orientation-Aware Location Based Mobile Services. Proceedings of the 3rd Symposium on LBS and TeleCartography, 2005.
12. Strachan, S., Eslambolchilar, P., Murray-Smith, R. gpsTunes – Controlling Navigation via Audio Feedback. Proc. of MobileHCI '05, September 19 – 22, 2005, Salzburg, Austria.
13. Wasinger, R., Stahl, C., Krüger, A. M3I in a Pedestrian Navigation & Exploration System. Proceedings of MobileHCI 2003, Udine, Italy, September 8-11, 2003.
14. Wonka, P., Wimmer, M., Zhou, K., Maierhofer, S., Hesina, G., Reshetov, A. Guided Visibility Sampling. ACM Transactions on Graphics, volume 25, issue 3, July 2006.

An Event Detection Service for Spatio-temporal Applications

WooChul Jung, DaeRyung Lee, WonIl Lee,
Stella Mitchell, and Jonathan Munson

Abstract. Sense-and-respond applications form an important class of pervasive-computing applications, serving domains such as driver services, field-force automation, and emergency services. The increasing ubiquity of networked data sources including sensors, mobile phones, online services, databases and data feeds presents novel opportunities for the timely use of the data, which is heterogeneous and available in huge quantities. When these data sources are embedded in the physical world, their location, possibly changing over time, becomes an important part of the context. Business applications will want to respond to the raw data from these sources in diverse ways, in a flexible and scalable manner. We introduce a spatio-temporal event detection service aimed at reducing the costs to application providers by enabling an infrastructure shared by many applications and subscribers. It offers a high-level spatio-temporal programming framework that enables application developers to more easily develop applications based on the sense-and-respond model.

Keywords: Sense-and-Respond, Spatio-Temporal, Event-based processing.

1 Introduction

An important class of mobile-computing applications involves sensing conditions in a mobile user's physical context and responding to those conditions, effectively in real time. This class is often known as "sense and respond." It is found in various domains, including mobile commerce, fleet logistics, and asset & personnel tracking.

We believe there may be substantial value in a general-purpose, shared, infrastructure that could support any and all of the above domains, simultaneously, over a large set of vehicles. Such an infrastructure would enable service providers to reach a broad customer base, without requiring an investment in their own infrastructure. With many services using the infrastructure, no one service must bear the entire load. We have developed the Spatio-Temporal Event Detection Environment (STEDE) to fulfill these requirements.

2 Related Work

As a framework for supporting sense-and-respond applications, STEDE is related to a broad array of work in real-time monitoring, event-driven systems, and context-aware computing. However, STEDE is focused on the spatial domain, initially in the context of automotive telematics. Furthermore, it does not address applications where

J.D. Carswell and T. Tezuka (Eds.): W2GIS 2006, LNCS 4295, pp. 22–30, 2006.
© Springer-Verlag Berlin Heidelberg 2006

extended and complex patterns of events are of interest. Space does not permit a full discussion of related technologies, but we briefly note those that are most closely related.

Chandy et al [4] describe an abstract programming model for dynamic applications that corresponds closely with the programming model of STEDE. The iSpheres Halo [9] platform offers a complete system for sense-and-respond programming. It does not, however, offer specific support for spatial events.

Some work addresses the specific application of location-based notification. Hightower et al [7] propose a layered software architecture for supporting location in pervasive computing. The Location Stack comprises seven layers, which have robust separation of concerns, and which provide the abstractions needed for location processing. The stack supports multiple sensing technologies at the bottom and multiple applications at the top.

Chen et al [**5**] describe a publish/subscribe system for spatial triggering, focusing on efficient matching algorithms. A set of spatial predicates and means of composing them are described by Bauer and Rothermel [2] and Nelson [10] presents a similar a set of spatial predicates, and extends this with some temporal operators.

Stronger support for event patterns is offered by ACT (Active Correlation Technology) [1]. ACT is a cross domain rule-based event correlation component developed by IBM. ACT rules use correlation, analysis and summarization patterns to transform low level input events into higher level complex events. The complex events can be used for notifications, alerts and to automatically trigger actions. Houdini [8] is a rule language and framework used for user-preferences policy management for telecommunication services. As such, it addresses a different set of applications than STEDE, and does not address spatial events.   Houdini and ACT runtime environments could both serve as a base rule engine for STEDE.

3   STEDE Concepts

Reflecting the structure of sense/respond applications, STEDE offers a rule-based programming model in which the application is partitioned in two parts: (1) a set of rules that operate on low-level input events and which, when triggered, produce high-level, application-defined events; and (2) logic that acts on the high-level events, which is deployed outside STEDE, and typically within the environment of the enterprise deploying the rule. [Figure 1] illustrates.

Applications subscribe to a rule to receive notifications of its triggering. The notification received includes any results that may have been computed in the evaluation of the rule. For example, a rule that determines if a mobile entity's position lies within any zones that have a promotion associated with them will, if the position is within any such zones, include the set of zones (via identifiers) in the notification to subscribed applications.

3.1   Rules

STEDE rules are essentially condition/action specifications, and may be as simple as a single Boolean expression, or a more complicated program with internal state. A rule condition is a logical expression (i.e., evaluates to true or false) composed of spatial and temporal logical functions joined through AND, OR, and NOT operators, and scalar functions joined with the usual relational operators. STEDE offers a number of built-in functions,. STEDE functions are described below.

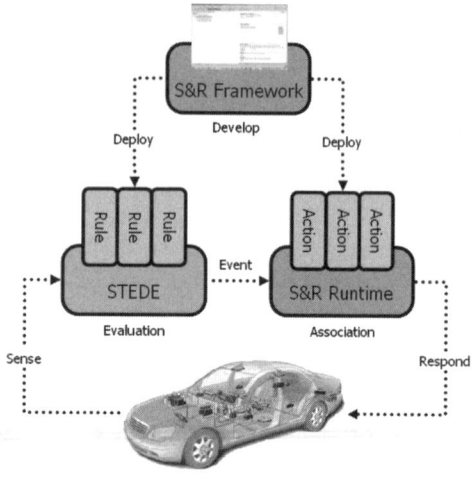

Rule inputs are modeled as continuously changing attributes of uniquely identified entities. In the telematics context, a vehicle is considered the entity, and its position, velocity, coolant temperature, etc. are its continuously changing attributes. Entities may also include static sensors, such as motion detectors, and "virtual" entities, such as an Internet address

Fig. 1. The S&R Framework

that supplies temperature at a particular location. A STEDE instance may make available to rules the attributes of multiple entity "spaces"—a set of vehicles, a set of warehouse sensors, and a traffic information service's set of virtual sensors.

Entity attributes (hereafter referred to as "inputs") are typed. A rule's input list is a list of typed input variables, much like a method's parameter list. Rule input lists do not specify the entities, or the sets of entities, which are to be the sources of the inputs. This specification is part of the rule subscription, described in Section 3.4.

3.2 Rule Functions

The spatial rule functions offered by STEDE operate implicitly on a subscriber position report, and have parameters associated with them that will also be input to the function. The *polygonID* and *pointID* parameters are identifiers for polygons and points. Polygons and points are specific kinds of rule resources, support for which is discussed in Section 3.5. **Table 1** below lists some of the built-in spatial functions of STEDE;

Table 1. Sample Built-in Spatial Functions

Name	Description
containedInPolygon	True iff the entity position is contained in the given polygon.
containedInPolygonSet	True iff the entity position is contained in one or more of the given set of polygons.
inProximityOfPoint	True iff the entity position is within the given distance from the given point.
inProximityOfPointSet	True iff the entity position is within the given distance from the given set of points.
distanceFromPoint	Returns distance (in meters) from the given entity position to the specified point.
distanceFromSubscriber	Returns distance (in meters) from the given entity position to the specified subscriber.

3.3 Rule Language

In the course of developing STEDE, we used multiple rule languages. Our first "language" was the construction of rules through building Java object structures of rule objects, much like an expression tree. We considered this to be not sufficiently easy to use. We then developed our own rule language and compiler. In the end, however, we realized that our primary value lay not in the rule language itself, but rather with the functions we provide outside the rule engine, such as rule optimization, application resource management, trigger reporting, and subscriber management (all discussed in later sections). Thus we adopted the approach of embedding existing rule engines within STEDE. This not only enables us to leverage the strengths of the particular rule engine in use, but gives us more flexibility in deployment as well.

The rule language we currently use is ABLE [3], an environment for building intelligent agents. ABLE supports different kinds of rule programming, through the variety of inference engines it provides. These include backward chaining, forward chaining, Rete networks, and simple sequential evaluation, among others. Below is an ABLE rule set that triggers when any one of a set of trucks is in a no-standing zone.

3.4 Rule Subscriptions

Subscriptions to rules exist independently of the rules themselves. Subscriptions include the following - (1) rule, (2) sources of the rule inputs, (3) the desired interval , (4) bindings for rule parameters , (5) subscription dates/times, (6) the destination to which rule-triggering events are to be sent, (7) the activation schedule for the rule, and (8) how long before the rule will "timeout" (remain untriggered).

3.5 Rule Resources

Simple rules may require only parameter values that are embedded in the rule itself. However, more complex rules should be able to refer to data that is persistent, and is managed outside the rule itself. The data may be referred to by multiple rules, and may be updated by processes outside the rule or rule engine. "Rule resources" are general-purpose storage locations for data. Some additional support is provided for those resources which are used in spatial functions, such as geometric points and polygons. For these resources, we support the notion of resource sets, to enable such functions as "containedInPolygonSet". Rule resources give much more flexibility to rules and enable rule programmers to write rules with respect to the resource, but not be concerned with managing the lifetime of the resource.

STEDE also supports rule resources that are associated with individual entities. This enables applications to maintain rule-accessible state for individual entities. For example, a rule may wish to record the time of each vehicle's entrance to a particular geographical area so that it can determine the total time the vehicle spent in the area.

Our rule subscription framework enables programmers to declare the resources that their rules require, enabling our system to ensure that such resources are present before the rule is deployed.

3.6 Rule Parameters

STEDE rules can be parameterized, which allows rules to be reused for different applications or subscriptions. For example, a rule that notifies if an entity's location has entered a certain zone and remained in the zone for a certain time period can be reused for multiple applications if the zone can be parameterized, and if the amount of time to stay in the zone can be parameterized. Given a zone around a dangerous area, the rule can be used for a safety-oriented application; given another around a shopping district, it can be used to deliver e-coupons.

3.7 Rule Example - Refinery Employee Safety

The state of Washington has recently imposed regulations affecting oil refineries, requiring them to adopt electronic safeguards for employee safety. An oil company that operates refineries there is now installing a pilot system based on an ultra wideband positioning system, which provides certain warnings based on employee positions. The refinery wishes to be notified in the following situations:

- An employee without the required safety training has entered a zone with particular safety concerns.
- A "fixed" asset has left a security zone.
- A person is in a work zone but has not moved in the last half-hour.
- A trainee is in a restricted area but is not in the company of a trainer.

Here are snippets of rules for the above situations:

```
ruleset CheckTrainingLevel {
...
: certLevel = (String)TEDS.getUserResourceValue(pos.getEntityID(), "certLevel");
: if (TEDS.containedInPolygonSet(pos, restrictZones, zoneList)) then {
    results.setTriggered(true);
    results.putResult("entityID", pos.getEntityID());
    results.putListResult("zones", zoneList);
  }
}
ruleset FixedAsset {
...
: assignZone = (String)TEDS.getUserResourceValue (pos.getEntityID(),"assignZone");
: if (!TEDS.containedInPolygon(pos, assignZone)) then {
    results.setTriggered(true);
    results.putResult("entityID", pos.getEntityID());
  }
}

ruleset TraineeWithSuper {
...
: restrictZone = (String)TEDS.getUserResourceValue(pos.getEntityID(), "restrictZone");
: if (TEDS.containedInPolygonSet(pos, restrictZones, zoneList)
      && TEDS.groupInProximity(pos, "Supervisors", 5.0) then {
    results.setTriggered(true);
    results.putResult("entityID", pos.getEntityID());
    results.putListResult("zones", zoneList);
  }
}
```

4 STEDE Application Development

STEDE supports four phases in the lifecycle of sense-and-respond applications: development, deployment, evaluation, and event-to-response association. [Figure 1] illustrates, in the context of automotive telematics. (The figure shows the response action as some communication back to the vehicle but that is a specific case; in general, the response action may be anything.)

Application development, deployment, and event-to-response association are supported by a set of tools we call the STEDE SDK and a set of libraries we call the S&R Runtime, which together we call the S&R Framework. The STEDE SDK supports rule and rule-subscription development, rule resource creation, response action development, and specification of event-to-response association.

The STEDE Runtime handles the deployment of rules, rule subscriptions, and rule resources, and evaluates the rules as input is received. The deployment process is described in Section 5, and rule evaluation is described in Section 5.3.

Events are sent to the destination specified in the rule subscription. The S&R Runtime, using the event-to-action map supplied by the application programmer, directs the event to the appropriate application-defined action. This component is not further described.

The STEDE SDK provides a set of Eclipse-based tools to assist in the creation of rules, rule subscriptions, and rule resources.

5 STEDE Runtime

The STEDE Runtime implements the various features of rule subscriptions. It installs the rule into the rule engine; it manages the activation schedule of the rule; it acquires the inputs specified in the subscriptions; it evaluates the rule when its inputs are received; and it forwards the event information to the address supplied in the rule subscription.

The architecture of the STEDE Runtime is shown in [Figure 2]. The Subscription Set Manager (SSM) and the Rule Subscription Manager (RSM) handle rule subscription deployment; the Data Acquisition Manager (DAM) and DAM Input Receiver handle data acquisition, the Rule Evaluation Manager (REM),

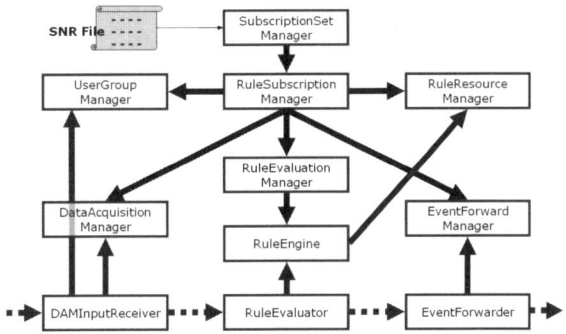

Rule Engine, and Rule Evaluator handle rule management and evaluation, and the Event

Fig. 2. STEDE Logical Architecture

Forwarding Manager (EFM) and Event Forwarder handle the process of forwarding events to applications (and, as we shall later discuss, other event consumers).

5.1 Subscription Deployment

The process of subscription deployment begins with the SSM unpacking the SAR file and interpreting the SNR file. Any user-group and rule-resource definitions are processed, and rule subscriptions are sent to the RSM for handling. The RSM maintains the set of rule subscriptions for the STEDE Runtime and manages rule activation and deactivation. For each rule subscription it requests the DAM to acquire inputs for the rule, it requests the REM to install the rule, and it passes the event-forwarding specification to the EFM for handling.

5.2 Data Acquisition

STEDE itself does not contain facilities to perform the acquisition of data, but it does define an interface that STEDE integrators will use to implement an input source for STEDE. An input source must supply a component that pushes inputs into the STEDE Runtime through a queue interface.

Once an input source is integrated with STEDE, developers can simply declare the input their rules require, as an *input-type*, *entity* pair, or an *input-type*, *entity-group* pair. All considerations of data acquisition mechanisms, including networks, data format, push vs. pull, and all other low-level input source characteristics, are hidden from the developer. To enable this, STEDE maintains a registry of data input sources, recording the input types they provide and entity spaces that they supply input from. The input source registry also records the characteristics of the input sources, such as whether the source pushes data into the rule system, or must be polled for it. If polled, the source describes how frequently the data changes, so that excessive polling can be avoided.

The primary function of the DAM Input Receiver is to identify which rules should be evaluated for the received input. We use an indexing scheme based on input type, entity, and entity-group. As rule subscriptions are received, they are added to the index based on their input specifiers. Our index is integrated with another data structure in the system called the Input Buffer, described below.

5.3 Rule Evaluation

When the Rule Evaluator receives input from the DAM Input Receiver, it first stores the input into a multi-level hash table called the Input Buffer. The Input Buffer has a first level of entries indexed by entity ID and entity-group ID. Each entity ID and each group ID referenced in a rule has an entry. Each of these entries is itself a hash table, indexed by input type name. Input Buffer entries are created when a rule subscription is received. At this time empty input-data objects are also created, in order to avoid excessive object creation during rule evaluation.

The Input Buffer also addresses two other issues. The first is that to avoid costly searches for data at evaluation time, a structure of shared pointers is used so that input data is only copied into the Input Buffer once, and rules access it from there directly, without further searching or copying. The second is that any item of input data may be referred to in a rule in multiple ways, because the entity that is the source of the data may be referred to by its ID or by a group to which that entity may belong. Therefore our Input Buffer provides a buffer location for not only an individual entity's data values, but also for each group known to the system. When data arrives, it is copied both to the entry for the individual entity it is from, but also to all groups to which the entity belongs.

The Input Buffer is also important in decoupling the arrival of input from rule evaluation. This decoupling enables us to trigger rule evaluation immediately, or to pace rule evaluation when input data arrives too frequently. It also allows us to manage the case where rules require inputs that arrive at different rates.

5.4 Event Forwarding

Rule subscriptions include an event-forwarding specifier that includes a destination "address" and a protocol indicator. Supported protocols include the common URL schemes such as http and file, class (for instantiating an application-defined class to handle the event), and an interface for supporting extensions.

6 Conclusions and Ongoing Work

We have described the Spatio-Temporal Event Detection Environment, a multi-application infrastructure that we believe can enable a wide range of event-driven mobile-computing services. STEDE's rule-based programming model provides a natural basis for support of the sense-and-respond programming paradigm, and its function set enables a range of spatio-temporal conditions to be expressed in a high-level manner. Its data-acquisition and rule-evaluation subsystems are designed for flexibility with respect to input sources as well as low latency in rule evaluation. STEDE's application developer tools support development-time, deploy-time, and run-time activities.

We are currently extending the STEDE work in two directions: continuing towards the ability to support large numbers of clients; and adding the ability to support rules with more complex temporal patterns.

References

1. Active Correlation Technology (ACT) (http://www-128.ibm.com/developerworks/autonomic/library/ac-acact/)
2. Bauer, M., Rothermel, K. Towards the Observation of Spatial Events in Distributed Location-Aware Systems. In Proceedings of the 22nd International Conference on Distributed Computing Systems Workshops, 2002.
3. Bigus, J. P., Schlosnagle, D. A., Pilgrim, J. R., Mills, W. N. III, Diao, Y. ABLE: A Toolkit for Building Multiagent Autonomic Systems. IBM Systems Journal, Vol. 41, No. 3, 2002.
4. Chandy, K.M., Aydemir, B.E., Karpilovsky, E.M., Zimmerman, D.M. Event-Driven Architectures for Distributed Crisis Management. Presented at the 15th IASTED International Conference on Parallel and Distributed Computing and Systems, November 2003.
5. Chen, X.Y., Chen, Y., Rao, F., An Efficient Spatial Publish/Subscribe System for Intelligent Location Based Services. Proceedings of Second International Workshop on Distributed Event-Based Systems, San Diego, 2003.

6. Graumann, D., Lara, W., Hightower, J., Borriello, G, "Real-world implementation of the Location Stack: The Universal Location Framework," in Proceedings of the 5th IEEE Workshop on Mobile Computing Systems & Applications (WMCSA 2003), pp. 122-128, Oct 2003.
7. Hightower, J., Brumitt, B., Borriello, G., "The Location Stack: A layered Model for Location in Ubiquitous Computing." In Proceedings of the 4th IEEE Workshop on Mobile Computing Systems & Applications (WMCSA 2002), pp. 22-28, June 2002
8. Hull, R., Kumar, B., Lieuwen, D., Patel-Schneider, P.F., Sahuguet, A., Varadarajan, S., Vyas, A. "Everything Personal, Not Just Business": Improving User Experience Through Rule-based Service Customization. In International Conference on Service Oriented Computing (ICSOC 2003), Rome, December, 2003.
9. iSpheres Corporation. Halo™. http://www/ispheres.com.
10. Nelson, G.J., Context-Aware and Location Systems. Ph.D. Thesis, University of Cambridge Computer Laboratory, 1998.

A Tourism Information System for Rural Areas Based on a Multi Platform Concept

Alexander Almer, Thomas Schnabel, Harald Stelzl, Jörg Stieg, and Patrick Luley

Institute of Digital Image Processing, JOANNEUM RESEARCH,
Wastiangasse 6, 8010 Graz, Austria
{alexander.almer, thomas.schnabel, harald.stelzl, joerg.stieg,
patrick.luley}@joanneum.at

Abstract. Tourism information is predominantly based on geographically related information and therefore, the tourism and leisure industries are currently searching for ways how to explore the potential of technologies for presenting geographical data. In this paper a concept and its realization for a multi-platform solution for a geo-multimedia tourism information system are briefly described. In general the system is targeting on two different user groups, the tourism boards as service providers and the tourists as end-users. The concept covers an efficient data management for the service providers and state of the art visualization techniques for online, offline and mobile solutions. These cover the 2D as well as the 3D visualization of geo related tourism information and also interfaces to third party platforms. Such platforms like Google Earth and Google Maps became even more important in the last year due to their extensive dissemination.

Keywords: tourism, information management, 2D/3D geo-visualization, mobile devices, multimedia, multiplatform, interactive maps, virtual database, Google Earth, Google Maps, virtual landscape.

1 Introduction

In general, the acceptance and the usage of geographical data have increased significantly in the last years, driven by free presentation services like Google Earth and Google Maps. 2D- and 3D- presentations of geographical data have become very popular and they are not only used by computer experts, but also by a broad community of Internet users. The challenge for the realization of an innovative concept is to bring two different views into one boat. On the one hand there are GIS specialists and representatives of the tourism boards aiming to offer and re-use existing geo-data from expert systems. On the other hand Internet users and tourists are interested to get geo-oriented data, like maps or points of interests about a special topic (hiking, biking, etc.) on multiple platforms to their specific needs. The main goal is to develop a methodology for combining these two approaches into a system.

Information interesting to tourists is location dependent by nature, meaning that thematic GIS systems can be used to offer this data in a location-aware way. The tourist's position then acts as a filter and parameter for system queries [7]. Advanced

J.D. Carswell and T. Tezuka (Eds.): W2GIS 2006, LNCS 4295, pp. 31–41, 2006.

visualization strategies will support a user friendly and geo-related information access. With the tremendous growth of the Web, the presentation of tourism products and services is a standard offering from the tourism and leisure industry. An innovative presentation of tourism information has to include the opportunity to present information on different output devices, has to consider the spatial context of the relevant data and to offer a customized, thematically and geographically information access [5]. Today and increasingly more in the future mobile devices such as mobile phones, PDAs and GPS devices are used by many people to get mobile access to information. Based on these requirements a technical concept for a multi-platform system has been developed targeting two user groups, the tourism boards as service providers and the tourists as end-users.

Basic developments were realized in frame of the EU-funded project ReGeo (Multimedia Geoinformation in Rural Areas with Eco-Toursim[1] – see also [4]). Further developments and the creation of the system were realized in close cooperation with the company MONASYS[2] and were funded from the Austrian Research Promotion Agency.

2   Technical Concept of a Multiplatform System

Multimedia presentation of tourist information signifies improved conveying of information by means of a well-aimed combination of different media, like text, sound, picture, video and animation. This presentation must not be restricted to one or two output media, but should work with every sort of existing technology like CD/DVD, Internet and mobile devices. The combination of digital elevation models (DEMs) with aerial and satellite images is the basis for the development of 3D views, virtual flights and panoramas. When combining this with tourist information in the form of texts, photos and videos a comprehensive information system emerges, which offers an easy access to information about the chosen region, tailored to the user's needs. This access is characterized by a high degree of interactivity, regarding the realization of multimedia CDs/DVDs as well as Internet solutions.

Most existing tourism information systems have clear shortcomings concerning geo-visualization (2D and 3D technologies), as well as in the management of data in a spatial context [2]. The acquisition and management of geo-multimedia (GMM) tourism information is a time- and cost-intensive process. The usage of aged information systems do not allow to gain the expected long term benefits of modern data management and presentation concepts.

To avoid such shortcomings the presented concept is based on a comprehensive GMM information pool with a centralized data storage and decentralized and personalized data administration which enables the provision of end-user applications on different output media such as World Wide Web, CD/DVD, info terminals and mobile devices. The problem of connecting to different databases and to have different requirements for the single platforms is solved using different web services (like data service for customized interfaces and a data packet generator). These services are used to provide the data different platforms concerning aspects like

[1] http://www.felis.uni-freiburg.de/regeo/
[2] http://www.monasys.org

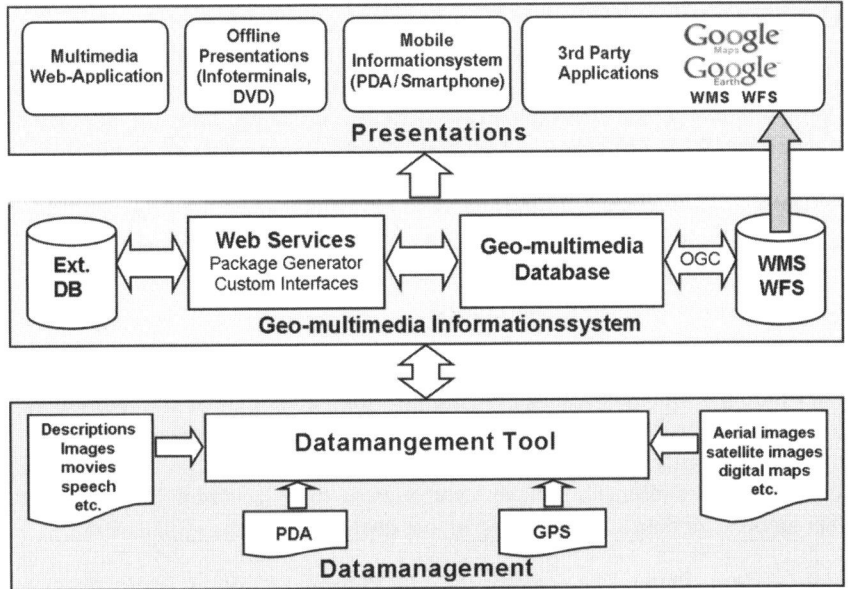

Fig. 1. Concept overview

different resolutions, qualities, etc. The basic elements for the complete system which considers the discussed requirements are shown in Figure 1.

Data management. A major module in the overall concept is a "Data Management Tool" (DMT) which has to meet the information providers' demands concerning usability and manageable data. This tool should allow the manual entering of data as an easily manageable method of building up a tourism database and of updating stored data as well as import and export capabilities from and to mobile devices (PDA, GPS) and other systems. A further important issue for such a DMT is to support the management of geographical oriented data (e.g. tracks, points of interests). To meet the requirements of different fields of application, the DMT must be flexible and easily adaptable to different scenarios and audiences.

Geo Multimedia Informationsystem. At this system architecture layer (Fig. 1) the different GMM data repositories like WMS, WFS, GMM database and also connections to external databases as well as services, which are providing interfaces to the GMM data, are implemented. The WMS, which is providing raster-data (digital maps), and the WFS, which is providing vector data (map overlays like polygons or symbols) are connected to the GMM database, which is merging the geo-data with multimedia files (like pictures or videos) using GIS functionalities of the database management system (DBMS). The web-services are dedicated to implement interfaces to the GMM data, which are used by the presentation- and data management layer application to query or store Geo-MM-data. A special web-service is the package generator, which is dedicated to generate content-packages for the use on mobile devices like a PocketPC or Smartphone (see mobile system).

Online presentation. The usage of the Internet offers an affordable, flexible and efficient exchange of information between guests and tourism organizations. With an eye to the design and the usability of the presentation the Internet offers a wide range of possibilities, which are not even partially used by most of the existing presentations. With increasing extent the Internet is used for travel planning and booking by prospective guests. Here, also digital maps play an important role, primarily in gathering general information but as well for detailed route planning [3]. The integration of 2D and 3D visualization technologies into existing web solutions and the connection with existing data bases allows a sophisticated presentation of tourism information in their spatial context.

Offline presentation. Unlike the online presentation, products on CD and DVD do not depend on an Internet connection and therefore are not suitable to present up-to-date information such as weather forecasts or events. Those products' strengths lie within the possibility to integrate high amounts of data that are up-to-date for a longer term (maps and multimedia content) and present those accordingly. The usage of higher amounts of data allows an even higher quality of the data visualization.

Mobile system (PDA). The combination of a PDA and a GPS-device is the state of the art choice to present and collect GMM data on the go and pursuing individual leisure activities. Therefore, a PDA data package can be generated contending all the necessary GMM data. This package can be pre-installed on the PDA for offline outdoor usage. For online usage of the PDA the data package contains only configurations of online data sources like WMS and WFS. Necessary GMM data is than downloaded on demand to the mobile device using a wireless data connection like UMTS or WLAN. Once downloaded data can optionally be stored locally on the PDA for further usage. Data gathered by the user can also be synchronized with the server on the go using the wireless data connection. This capability is useful, if this data should be made available for other mobile users immediately.

3rd party applications. Existing visualization platforms can be used to provide tourism information over different existing channels. Specific interfaces to the used platforms have to be created to allow an enhanced dissemination of the data on established and well known platforms to a broad community. Platforms like Google Earth and Google Maps have to be considered due to there high acceptance in the Internet community.

3   Realization of the Concept

3.1 Content of a Tourism Information System

Interactive maps are created on the basis of aerial and satellite images as well as of existing maps to allow a space related presentation of the thematically tourism information. For the system realization within the area "Alpenregion Bludenz" in Austria, maps of different scales and orthophotos have been used (see Figure 2). Furthermore, a digital elevation model was available from the cost free SRTM

(Shuttle Radar Topography Mission[3]) data with a pixel resolution of 90 meter which is sufficient for 3D-landscape-visualization. Focusing on relevant tourism themes of the region the following data are relevant:

- Tours and tour points. the coordinates of a hiking or biking tour, the most important waypoints, refuges and viewpoints along the route.
- Tourist infrastructure: info points, public facilities, sports facilities, etc.

Fig. 2. Orthophoto and city map

- Places of interest: castles, palaces, museums, excavations, etc.

3.2 Data Management

The data-management-tool (DMT) was designed as a user-friendly, easily adaptable tool in order to provide all essential functionalities to fulfill the task of managing the contents of a tourism information system. Further, this tool allows the customer (data provider) to view and manipulate different multimedia information of point- (hotels, sights, etc.) and vector-objects (e.g. hiking and biking tours). The DMT (Figure 3) bears the advantage that it is completely customizable. It can be configured using local configurations or by a central server. This enables multiple users to use different profiles which include configurations concerning the manageable themes and areas. The data itself is stored in a centralized database with the ability to be used by multiple clients simultaneously. The three main modules of the DMT are:

- The data-import/export module (interfaces for connecting to mobile devices and other databases/systems)
- The data-visualization module including the 2D presentation with different information layers
- The data-manipulation module includes tools for adding and editing GMM data as well as simple GIS functions

The data import and export module provides interfaces to mobile devices, GPS devices and external databases and systems. So, existing data can be re-used. Furthermore, data acquired via a mobile device can be integrated into the central database. In addition, the user may export existing data to a PDA, and tracks as well as points can be exported to a GPS device or any other system using a defined XML structure. In order to be more flexible the DMT uses XML files to define the structure of the used data. With the implemented interfaces it is possible to connect to other external databases and exchange data easily. The visualization module allows the user to view the available tourism information on two dimensional maps. The different themes are visualized by representative symbols which are also assigned through the configuration file. Next to that an intuitive navigation on the map (pan, zoom, switch maps, etc.) is offered by the DMT. Another important module is the data-manipulation module that allows a user-friendly way to add new data into the

---

[3] http://www2.jpl.nasa.gov/srtm/cbanddataproducts.html

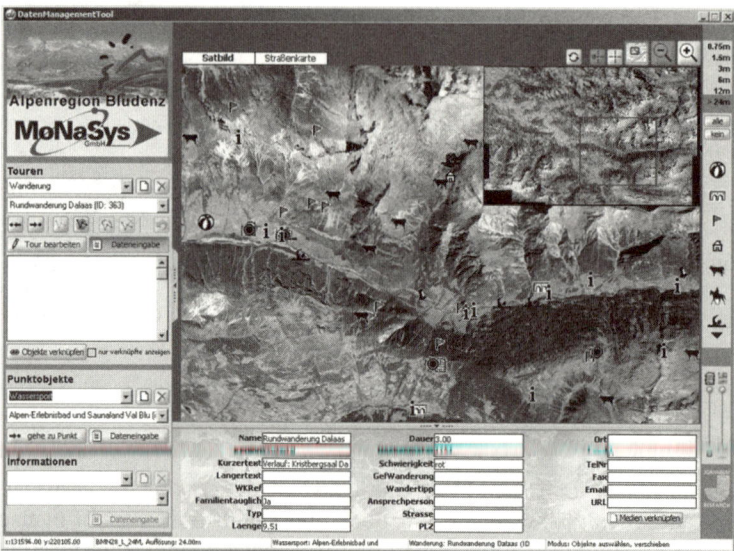

Fig. 3. Data Management Tool

system and to manage the related multimedia information. Besides that, objects (points, tracks, areas) can be digitized on the map and in addition it is possible to associate points with specific tracks.

3.3 Online Presentation

The main goal of the online presentation is to present information about the regional tourist infrastructure and leisure themes in their spatial context. To meet the different needs of the users, it offers two different approaches to the information. On the one hand there is a theme based access that gives the user a structured approach, on the other hand it is also possible to access all information via a geographical approach. Nevertheless, the user is able to switch between both and also combine these two approaches whenever needed to get all information in the way he prefers. In addition, the online presentation offers downloadable packages which are generated within the DMT and include data intended to be used on either a PDA or a GPS device. Through this distribution channel the tourism region can offer a public service for location aware data information. Also this system can be seen as an additional module to existing presentations. It can be integrated into existing solutions which are used for a wider range of usage like booking systems, CRM systems etc. The integration has been realized using frame technology and links between existing and new geo-content have been established.

Theme based approach. The theme based access starts with a general information about the region. At this point the user can select a specific theme (e.g. hiking or biking tours, leisure activities, events) from a menu and gets an overview about this chosen topic and a list of the available tours and point-objects. After choosing a specific object, additional information which includes a detailed description, contact

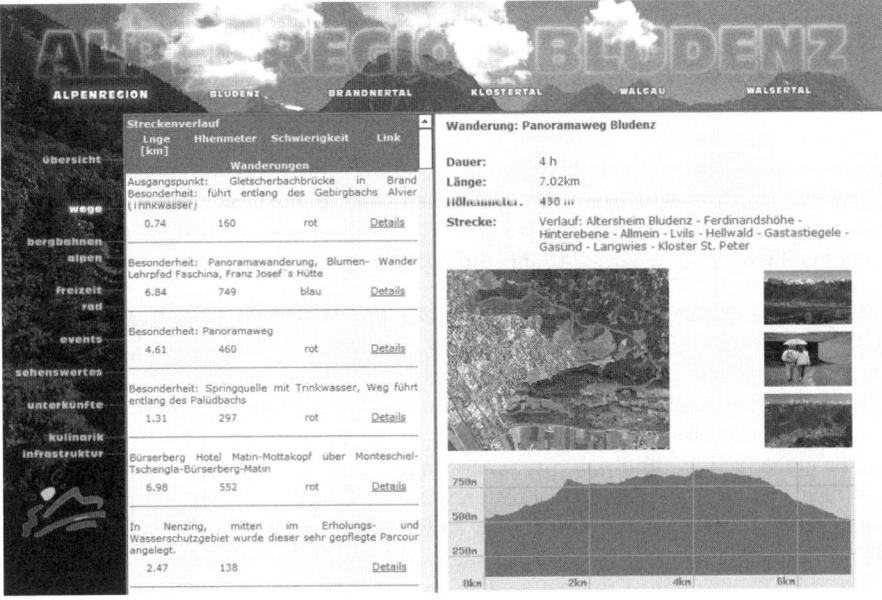

Fig. 4. Online theme based approach

information, different attributes and the position of the object are offered. For hiking and biking tours also a height profile is provided. Figure 4 shows an online visualization example of this approach.

In addition to the information of the local data-base data from other systems or data providers can be integrated into the presentation. This can be accomplished by simple links or embedding the information into frames.

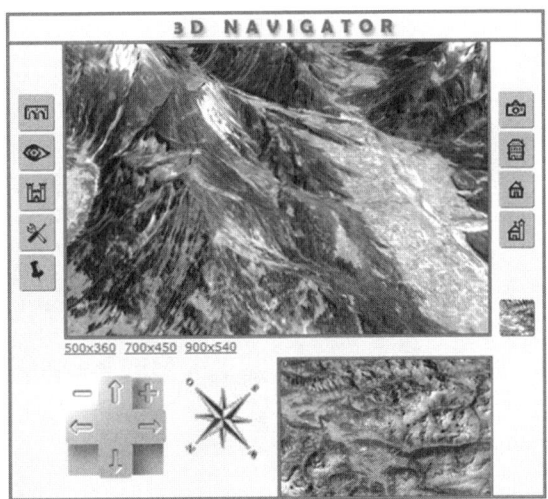

Fig. 5. Online 3D navigator

2D and 3D visualization. Next to the theme based approach, also 2D and 3D-visualization shows geo-oriented tourism information such as biking, hiking tours or sights in their spatial context has been developed. The shown data has been integrated by the usage of the DMT. The tours are visualized with lines and the infrastructure is shown by means of points and symbols. Thus, the user may choose information he really requires, either by selecting particular themes or by gathering information in a geographical oriented search. The desired information is presented immediately in its geographical context. In contrast to the 2D-view, the 3D-view of a region shows also the terrain elevation and therefore gives a realistic impression of the landscape.

A real time 3D-model as shown in Figure 5 is realized on the basis of a digital elevation model from which the relevant elevation data are taken. This technology of visualization offers the user an easy interaction method. He is able to zoom to an area, twist the model and change his perspective. Furthermore, also flight paths can be predefined which allow virtual flights through a tourist region [6].

Interfaces to existing visualization platforms and services – Google. Google Earth (GE)[4] and Google Maps (GM) became very popular in the last year. GE is a free-of-charge, downloadable virtual globe program that shows the entire earth in 2D and 3D using available satellite images, aerial photographs and GIS-data. GM on the other hand is a free, web map server application and technology provided by Google that can be embedded on third-party websites via the GM API[5] (application programming interface). Because of a huge market presence of Google, 2D and 3D platforms are much more known by the public than some years ago. So, a tourism information system should also allow integrating their data into such a platform. GE and GM allow this, using defined interfaces. GE used the Keyhole Markup Language[6] that allows the direct integration of point, vector and raster objects into the 3D visualization application. GM on the other hand can be integrated into the online presentation itself using the GM API. A direct connection to the GMM database of the tourism region can also be enabled using this API in connection with server side scripts like PHP.

The following figures show the integration of the same data as already shown above into GE and GM.

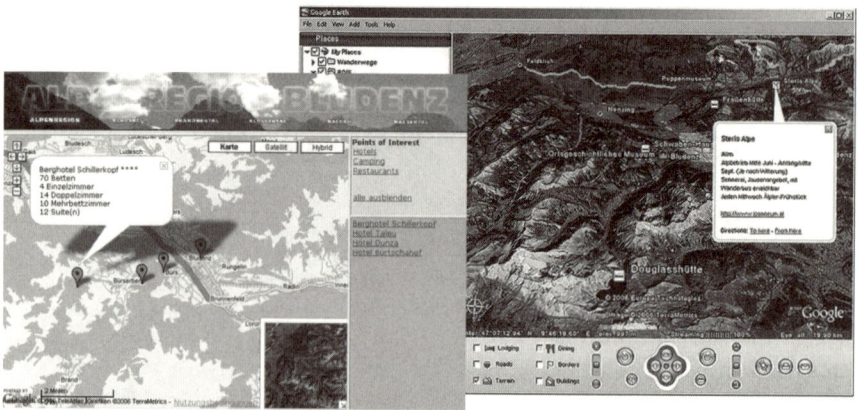

Fig. 6 & 7. Data integration into Google Earth and Google Maps

---

[4] http://earth.google.com/
[5] http://www.google.com/apis/maps/
[6] http://earth.google.com/kml/kml_intro.html

To be noticed here is the fact we have a direct interface to the GMM database. This allows the presentation of all data, which has been entered using the DMT (see chapter 3.2). Every modification is directly shown on this platform and ensures that the shown information is always up-to-date.

3.4 Offline Presentation

The offline application offers the functionality of a digital brochure with the advantages of using a digital media. In principal the content can be compared with the online version but there is a slightly different approach to the information access for the user. Here, the primary approach is via a geographical selection. In a second step the user is able to select vectors and points and to request additional

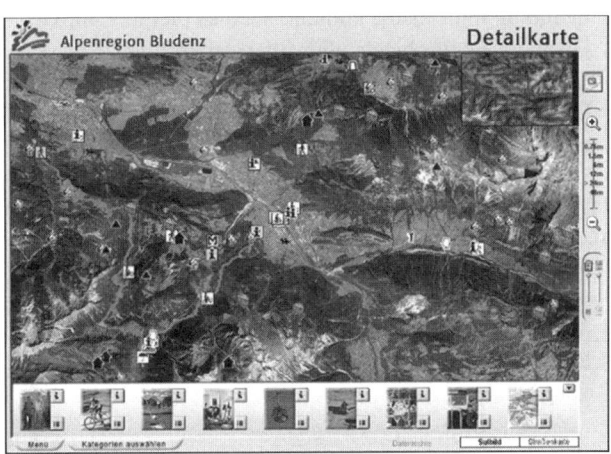

Fig. 8. Offline application

multimedia information to the object. Figure 8 shows the developed offline solution with the shown description of a hiking trail. The digital brochure can be used as a cost-efficient alternative to the printed version of the brochure. Also it can be used in info-terminals. This does not only have the advantage to show multimedia content in a sophisticated way but also enables the tourism region to frequently update the data on the CD/DVD based on the existing GMM data with hardly any financial expenses.

3.5 Mobile System

The Mobile Tourism Information System was developed for a PDA running "Microsoft Pocket PC 2003" or higher in combination with a GPS functionality. These devices are high performance mobile multimedia computers with high resolution color displays. In addition, a memory card was used to expand the internal memory of the PDA and store cartographic as well as object data [1].

The used basic data is generated using the DMT to get the data from the GMM database. Up-to-date information can be added by exporting user-selectable regions from the DMT or by installing downloaded packages from the Internet. Those packages can be configured online, giving the user the chance to define which maps, regions and types of object data he wants to download, depending on his needs, bandwidth and available storage space on his PDA.

As mentioned before, the Mobile Tourism Information System on the PDA can be divided into the presentation- and the data acquisition-part.

The presentation on this device (Figure 9) includes the following information levels and functionalities:

- General information about and impressions of the region using multimedia data.
- Display of different types of maps stored as tiles with the opportunity of interactive navigation and zooming.
- Visualization of objects, e.g. tours and hotels, on the map and their details including multimedia content, such as pictures, movies and sounds.
- Display lists of objects with filter-capabilities.
- Interactive selection of different layers for the spatial presentation.
- Visualization of the current GPS-position if available and display objects within a user-defined distance.

Data acquisition is realized by the following functionalities:

- Record tours and points via GPS while being in the field and link photos or voice notes recorded with the PDA to those objects.
- Keep a digital, location aware, personal diary of your holidays.

With the focus on user-friendliness and performance, the Mobile Tourism Information System aims at all state-of-the-art PDAs including higher-resolution devices with VGA resolution (480x640 in contrast to 240x320 pixels). Despite the higher amount of image data displayed on such a higher-resolution device in comparison to a regular PDA, the performance while navigating on the map is optimized. This is achieved by using different types of maps depending on the device's resolution and tiles of a different size.

Fig. 9. Mobile device

4   Conclusions

The paper introduces an innovative and user-friendly concept for a geo-oriented management and presentation of tourism information on different output devices considering the spatial context of the data. Advanced 2D and 3D visualization technologies support the geographic oriented information access and allow an optimized orientation and navigation within a virtual landscape of tourism regions. The data management system, tailored to the service provider's requirements, allows a time and cost efficient administration and actualization of the information. Also, it enables the usage of data from a virtual network using interfaces to existing systems. The system allows a tourism organization as the content provider to manage their data completely themselves which includes new as well as existing data. This allows to keep the quality of the relevant information on a high level and guarantees usable information for sophisticated presentations as described in this paper.

Location awareness, a multi platform data presentation using innovative visualization technologies and personalization of the access to information are more and more important prerequisites for the acceptance of tourism information systems. This also includes the high demand of integrating geographical aspects into tourism information platforms, which is an important ongoing issue for the tourism sector.

References

1. Almer A., Luley P., Nischelwitzer A. (2003): Location Based Tourism Information Systems on Mobile Multimedia Devices. GNSS2003 Global Navigation Satellite System – Congress – Graz, 22-25 April 2003
2. Almer A., Stelzl H. (2002): Multimedia Visualization of Geoinformation for Tourism Regions based on Remote Sensing Data. ISPRS - Technical Commission IV/6, ISPRS Congress Ottawa, 8-12 July 2002.
3. Faby, H. (2004). Individuelle Reisvorbereitung mit Internetkarten: Status quo und Potenziale. In: Kartographische Nachrichten, Heft 1, 54. Jg., 3-9.
4. Frech, I., Koch, B. (2003): Multimedia Geoinformation in Rural Areas with with Eco-Tourism: the ReGeo System. ENTER 2003. 29 to 31 of January in Helsinki, Finnland. 2003.
5. Raggam K. and Almer A. (2005): Acceptance of geo-multimedia applications in Austrian tourism organisations. ENTER 2005. 26 to 28 of January in Innsbruck, Austria. 2005.
6. Schnabel, T., Almer A., Stelzl H. (2004): Advanced Visualization Techniques for Rural Areas. Information Systems for Agriculture, Forestry and Rural Areas, Seč u Chrudimi, 19.-21. April 2004.
7. Zipf, A. (2002): Adaptive context-aware mobility support for tourists. IEEE Intelligent Systems, Volume 17, Number 6, November/December 2002; Intelligent Systems for Tourism. p53-64.

Prediction of GPS Multipath Effect Using LiDAR Digital Surface Models and Building Footprints

Jing Li[1], George Taylor[2], David Kidner[2], and Mark Ware[2,*]

[1] Department of Geography, University of Leicester, London Road,
Leicester, LE1 7RU , England, UK
jli@glam.ac.uk
[2] GIS Research Centre, Faculty of Advanced Technology,
University of Glamorgan, CF37 1DL, Wales, UK
getaylor@glam.ac.uk, dbkidner@glam.ac.uk, jmware@glam.ac.uk

Abstract. This paper aims to investigate how 1m LiDAR data and 2D building footprints can be used to predict GPS multipath effects in urban areas. A ray tracing model is implemented in order to model reflected and diffracted GPS signals. Some preliminary results are presented and explained in detail.

Keywords: LiDAR, GPS, multipath, Digital Surface Models (DSMs).

1 Introduction

Previous research findings have shown that LiDAR DSMs can model the line of sight (LOS) between the GPS satellites and receiver points on the ground at a very high accuracy [1]. It is anticipated that 1m spacing LiDAR DSMs can also be used to predict GPS multipath effect in order to assess the positioning accuracy of conventional and high sensitivity GPS receivers for location based services, as there is a growing demand for the use of GPS and/or the future Galileo system for transportation applications such as positioning-based road user charging and bus positioning [1][2]. High sensitivity GPS receivers are currently under development due to the fact that many receivers have to operate in the lower signal to noise regimes found in personal, hand-held, mobile and asset tracking applications and are consequently more prone to multipath errors [3].

In practice, both GPS code and phase observations are liable to multipath. Multipath effect is highly dependent on the surrounding geographic features near a GPS receiver, and therefore can not be removed by differential GPS which removes most of the errors in GPS positioning. The term multipath is derived from the fact that a GPS signal might follow several paths to a receiver's antenna. The signal can be reflected from buildings or the ground and create a range error of several metres or more in C/A code measurements [4]. As illustrated in Fig.1, apart from the direct ray, a GPS antenna may receive diffracted and reflected rays which can deteriorate the positioning accuracy.

---

* Corresponding author.

J.D. Carswell and T. Tezuka (Eds.): W2GIS 2006, LNCS 4295, pp. 42–53, 2006.
© Springer-Verlag Berlin Heidelberg 2006

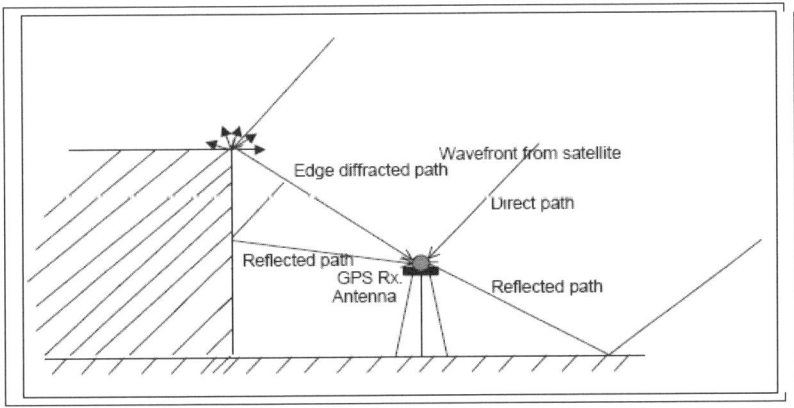

Fig. 1. Typical multipath scenario [5]

2 Methodology and Data

The methodology used in this study is known as ray tracing and is an approach to 3D deterministic radio wave propagation modelling, it is also commonly used for surface-rendering. The prime motivation for this study is that highly accurate LiDAR DSMs at 1m or even higher 25cm resolution have been made available in recent years, 2D building footprints in digital form are also undergoing an improvement in terms of positional accuracy and topological relationship in the UK [6] Therefore, it is worthwhile to develop a test-bed simulator specially tailored to these two GIS data sets. Fig. 2 shows an overlay of building footprints with a 1m spacing LiDAR raster DSM covering part of London Bus Route No2. The image on the right is the same LiDAR model without the building footprints

Fig. 2. 1m LiDAR raster DSM and 2D building footprints covering part of London Bus Route No2

As seen in Fig.2, the image on the left shows that the 2D building footprints closely line up with the 1m LiDAR raster DSM. Moreover, urban features can be

44   J. Li et al.

clearly recognized from the 1m LiDAR DSM even without the 2D building footprints (i.e. the image on the right).

GPS signal operates at UHF band (i.e. L1 1575.42MHz, L2 1227.6MHz). It is generally accepted that the wavelengths (L1:19cm and L2:24cm) are small in comparison with the dimensions of the obstacles, and the building walls are treated as specular reflectors. It is well known that ray tracing techniques are based on the same assumptions.

3 Diffraction

Signal diffraction may refer to the fact that the signal is partially masked or unmasked and can still be tracked by the receiver. For example, it is possible that GPS signals bend around the edges of a wall although from the geometrical point of view the satellites are obstructed [7]. As a result, the diffracted signals are weakened and travel a longer distance compared to the direct signals which consequently introduce a range error into positioning. Often, the Signal-to-Noise ratio (SNR) is highly correlated with signals contaminated by diffraction. Low SNR may indicate the occurrence of diffraction especially when the receiver is placed close to a building.

3.1 The Implementation of the Single Knife-Edge Model in a 3D GIS

In this study, a simple single knife-edge diffraction model is first implemented according to the International Telecommunication Union Radio Communication (ITU-R, 1999) standard. Such a model is illustrated in Fig 3 and Fig 4 which account for both LOS and Non Line-Of-Sight (NLOS) cases.

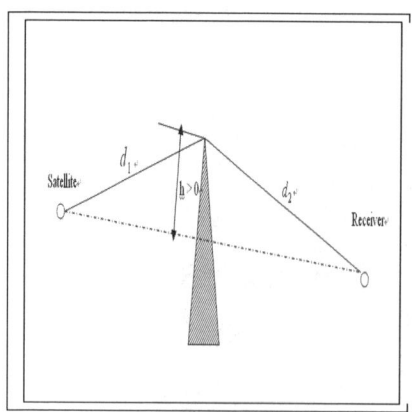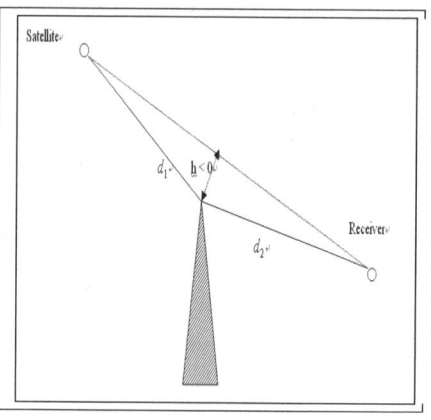

Fig. 3. (a) NLOS   (b) LOS

Where:
    h – effective height of the obstruct (i.e. perpendicular distance to the LOS)
    d1 – distance of the obstruct from the satellite
    d2 – distance of the obstruct from the receiver

For v greater than -0.7, the diffraction gain G(v) can be approximated using equation (1)

$$G(v) = 6.9 + 2\log\left(\sqrt{(v-0.1)^2 + 1} + v - 0.1\right) dB \tag{1}$$

Where the Fresnel parameter $v = h\sqrt{\dfrac{2}{\lambda}\left(\dfrac{1}{d_1} + \dfrac{1}{d_2}\right)}$ $\lambda = 0.19$m for GPS L1

Note that v is negative, when there is clear LOS between the satellite and receiver (i.e. h is negative).In that situation, the attenuation is less severe compared to the NLOS signal with a positive v value (See Fig. 4).

For the NLOS signal (i.e. v greater than 0), the diffracted signal gets increasingly weaker as the value of v increases. This trend is clearly depicted in Fig. 4 that is computed using equation (1).It is quite clear from Fig. 4 that the diffracted signal is weakened, and may not be tracked by the receiver if the loss is too high.

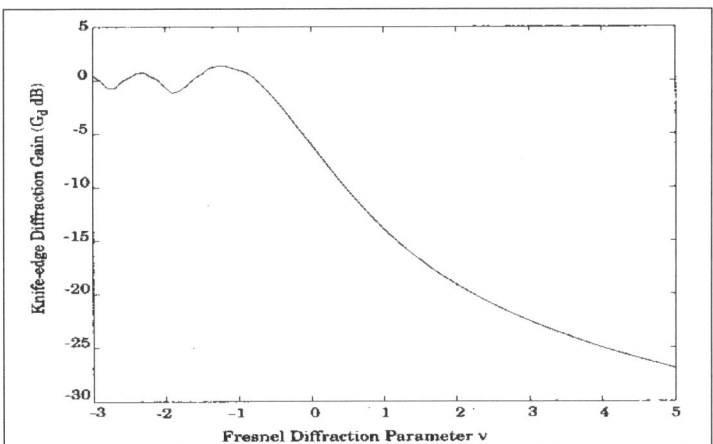

Fig. 4. Knife-edge diffraction gain as a function of Fresnel Diffraction Parameter v

The single knife-edge diffraction model is implemented in ArcGIS9 3D analyst as follows:

For the LOS signals, the shortest perpendicular distance from the LOS vector to the terrain profile (i.e. h less than 0 in Fig 3 (b)) is computed so that the 3D coordinates of the terrain point closest to the LOS vector is chosen as the obstruction point and then used to calculate the Fresnel parameter v. In this case, the LOS vector is above the terrain profile.

In terms of the NLOS signals, the LOS vector is below certain portions of the terrain profile, so the coordinates of the obstruction point can be found by interpolation along the terrain profile. This approach may be considered as finding the nearest edge point to the receiver provided that the building features are adequately modelled in the 1m raster LiDAR DSM. As stated earlier, it is generally accepted that, for GPS signals, the buildings closest to the receiver are more likely to contribute to the propagation process.

3.2 Multipath Data Collection and Results

In order to examine the effect of the single knife-edge model on the positioning accuracy, a pair of Leica Geodetic-grade GPS receivers with AT502 dual-frequency antennas were used to collect data that is expected to be affected by the diffracted signals. It should be noted that multipath mitigation techniques have been implemented within the Leica receivers [8], which are designed to reject long-delay multipath signals. However, multipath mitigation schemes are often manufacturer specific and assumed to operate in conditions with high signal to noise ratios [3]. In this study, the GPS data were collected at the relatively lower signal to noise ratios (i.e. 33-42dB) than the expected ones (40-48dB) at a range of satellite elevation angles, which often imply that the direct LOS signal is not present, and the resultant positional error is large.

The simulation is carried out in a 3D GIS environment. Data were collected over a period of 30 minutes in front of a building on the campus of the University of Glamorgan. In this period, SV10 was found to be affected by diffraction. As shown in Fig. 5, the solid green lines indicate the visible portions of the LOS between the receiver and the satellite, the invisible part of the LOS is indicated by the solid red lines. The reflected rays, indicated by the dotted green lines, are also present over the same period. The modelling of the reflected rays will be described in the next section.

It is very clear from studying Fig. 5 that SV10 was moving from being invisible to visible to the receiver over a period of 30 minutes, as the elevation angle increases. The satellite positions are computed with the broadcast orbits which can be extracted from Rinex navigation file (Receiver Independent Exchange Format).

Fig. 5. The screenshot showing SV10 moves from being invisible to visible to the receiver over a period of 30 minutes

Fig. 6 is a plot of the computed elevation angles of SV10 against the single knife-edge diffraction gain expressed in decibel over a period of 30 minutes. The gain is estimated using equation 1 for both LOS and NLOS signals. As expected, the signal gets stronger, as the elevation angles increases. It is quite evident from Fig 5 and 6 that 1m spacing LiDAR data successfully identified this transition period in which the direct signal is partially masked and unmasked. Note that the fractional parts of the elevation angles on the x axis are omitted in Fig. 6.

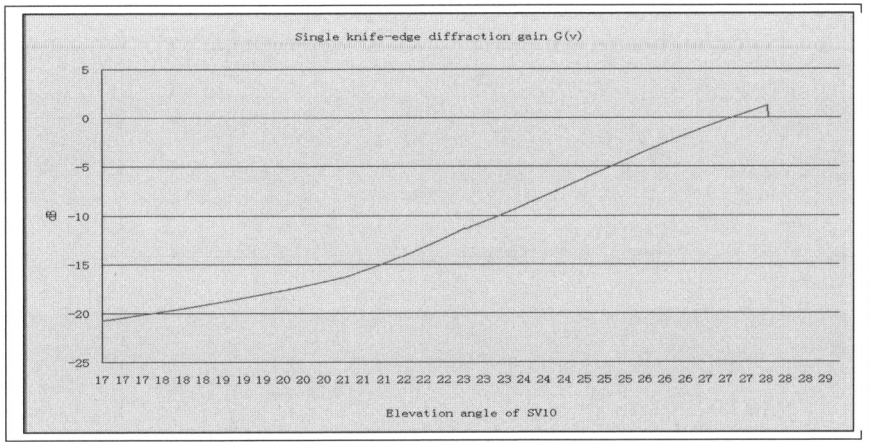

Fig. 6. Elevation angle of SV10 VS the single knife-edge diffraction gain for point 1

Having identified the affected satellite and the transition period, the Double Differencing (DD) residuals are used to give an indication of the magnitude of the possible pseudorange multipath error. This is because the base and rover stations are placed so close together (about 100 metres apart) on campus, almost all orbit errors, receiver clock errors, atmospheric effects are eliminated. Therefore, the DD residuals are mainly comprised of multipath error and a small amount of measurement noise.

During the data collection, one station, whose coordinates were known to be accurate to a few centimetres, was set up as the base station on the roof of the School of Computing, thus estimating the baseline vector between the rover and base station is equivalent to estimating the coordinates of the rover station that is expected to be contaminated by multipath. The DD residuals are the differences between the actual observed double difference measurements and those computed by the least-squares solution for the baseline vector. The commonly used approach to double differencing is known as "weighted least squares". As described in [9], the solution is given by Equation (2):

$$\hat{x} = \left(A^{T}WA\right)^{-1} A^{T}Wb \qquad (2)$$

Where W is the data weight matrix, b is a vector containing the double-differenced residual observations. A is the design matrix. The weight matrix W for double-differenced data is often constructed using " the law of error propagation". A detailed explanation of how the design and weight matrices are derived can be found in [9].

Fig.7 shows the code double-differenced residuals of SV10. It is very clear from Fig.7 that very large residuals with the maximum up to 30 metres can be observed on

SV10 which means that the effect of diffraction on SV 10 is more significant compared with those satellites unaffected. This agrees with the prediction result shown in Fig.5and 6.

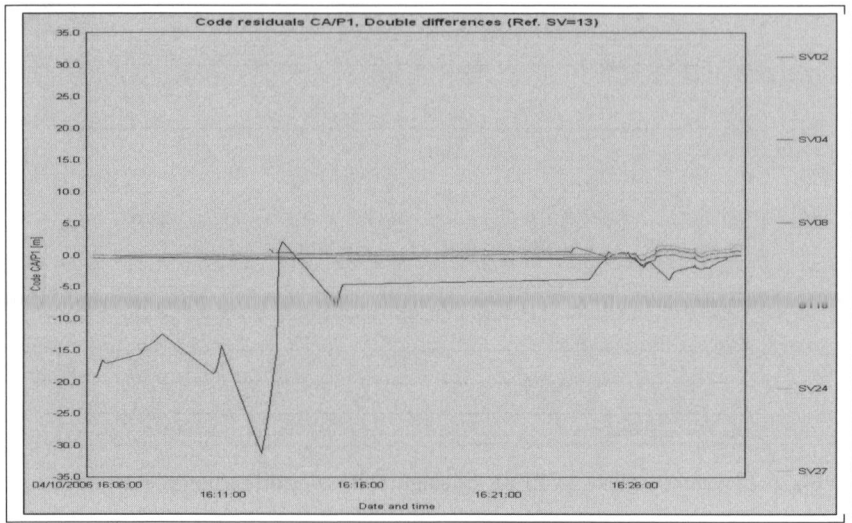

Fig. 7. Double differencing residuals for Point 1

To sum up, by applying the single knife-edge model to 1m LiDAR data, the effect of single knife edge diffraction can be predicted successfully, the signals weakened by the obstructions are tracked by the receiver and is the major cause of the large DD residuals. Note that similar results were obtained on two other locations on campus. However, due to the page limit, these results are not presented in this paper.

4 London Bus Data

The experiment was carried out along part of London bus route No2 over a 5 minute period. A low cost single GPS receiver was mounted on the top of a bus whose height is 4.4m. In Fig.8, the bus is travelling from north to south. It is quite evident that removing SV15 from the single-epoch least square solution brings an improvement of around 40m. The GPS points are brought back on to the road centreline at the beginning and end of the road without using SV15. The elevation angle of SV15 is just above 15 degrees, and introduced large positional errors. Note that other error sources such as atmospheric effects and satellite clock can not cause such severe positional errors. Similar improvements as such are observed on other part of the route as well at different times.

In order to produce a visibility map for SV15, the antenna height of 4.4m is added to the height of each pixel cell centre on the 1m raster LiDAR DSM, the LOS analysis is subsequently performed at each cell centre. Therefore, the resultant visibility map has a very high resolution of 1m, and is displayed as a backdrop in Fig.8. It can be seen from Fig.8 that SV15 is frequently masked and unmasked by the buildings especially in the two transition zones at the beginning and end of the road where SV15 becomes invisible to visible. Also, SV15 is not visible most of the time when travelling along the middle part of the road due to the urban canyon effects. However, we do not have sufficient evidence to say that this large positional error is due to multipath although it is very likely.

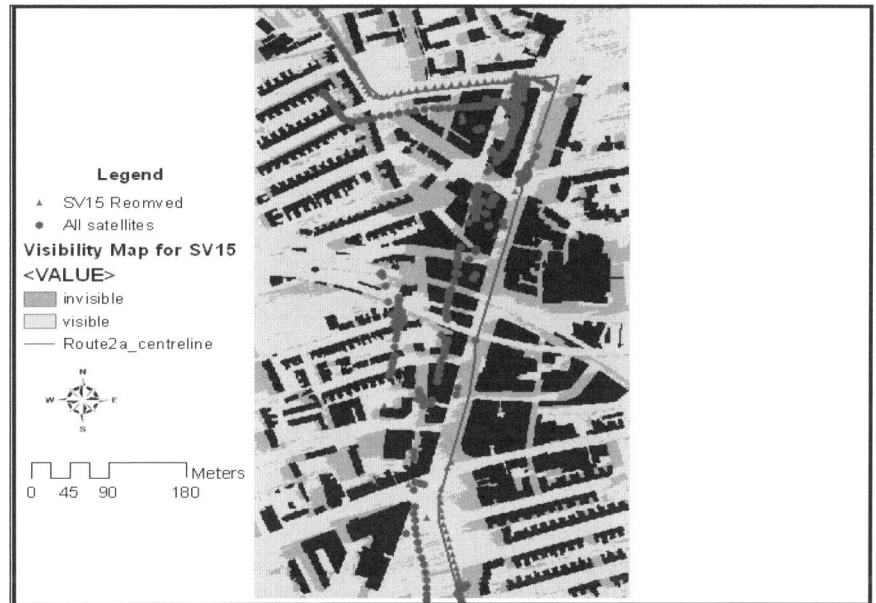

Fig. 8. Improvements without SV15

5   Reflection: Simulations and Results

5.1   Ray Tracing for Reflected Rays

In general, ray tracing techniques can be classified into two main groups: techniques based on the shooting and bouncing rays method and techniques based on the image method [10]. Image-based ray tracing appears to have some advantages. Instead of using the 'brute force' approach of launching many rays at very similar angles, the technique considers all obstructions as potential reflectors and calculates their effect using the image method [11]. The image method is used in this study (Fig. 9). As shown in Fig.9, it appears that the signal is emitted directly from the image source (I) given an original signal source (S) and a building segment. The image point (I) is symmetrical to the signal source (S) with respect to the building segment. Note that two additional tests have to be made before a reflected ray is successfully identified. First, the reflection point (R) must be located onto the building segment in order to create a reflected ray. The second test is to check if there are any obstructions between the signal source (S) and the reflection point (R) which block the LOS. The details of the implementation of image-based ray tracing algorithm are presented as follows:

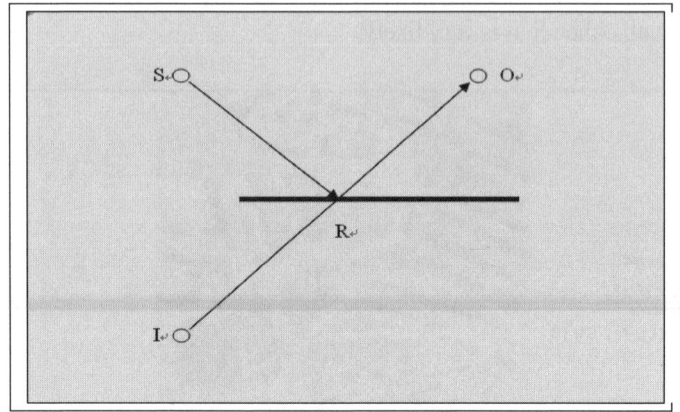

Fig. 9. Application of the image method to the ray reflections

1. Calculate the 2D coordinates of the image point with respect to the current building segment in the scene.
2. Calculate the 2D coordinates of the intersection point (R) between the current building segment and the line connecting the image point and the receiver point (O). The calculation of the intersection point is completed in 2D, which requires the 2D building footprints. The heights of both image and receiver point are set to zero in this step.
3. Perform the 3D LOS analysis between the current image point and the receiver point using 1m spacing LiDAR raster DSM. The reflection point exists only if the obstruction point closest to the receiver is identical to the intersection point computed in Step 2. The height of the obstruction point is interpolated along the LOS, which is also the height of the reflection point.
4. Perform the LOS analysis between the reflection point and the satellite. If unobstructed, a reflected ray is detected.

Steps 1-4 are performed iteratively for each of the building segments in the scene and the satellites in the sky above a particular elevation mask angle (e.g. 15 degrees).

As illustrated in Fig.10, dotted green lines are reflected rays. It should be noted that the edge effects of the 1m LiDAR DSM must be taken into account when performing the LOS between the reflection point and the satellite. This is because the LOS may be blocked by an obstruction point on the building wall that is very close to the reflection point (e.g. only a few centimetres away). In such a circumstance, the satellite should be deemed visible rather than invisible. Furthermore, the developed simulator can perform very fast, for example, reflected and diffracted rays can be identified at one location within one or two seconds.

Fig. 10. 3D visualisations of reflected and direct rays

5.2 GPS Receiver Correlator

To date, various mulipath mitigation techniques have been built into GPS receivers such that long-delays can be minimized. However, short delays (e.g. less than 30m) are still needed to be dealt with.

The heart of a GPS receiver is the autocorrelation function depicted in equation (3).

$$R(\tau) = \frac{1}{1023} \int_{0}^{1023} p(t)p(t+\tau)d\tau \qquad (3)$$

Where: R is the autocorrelation function,

τ is the lag

P(t) is the prn code at time t (either plus or minus 1)

In principle, the incoming code is correlated with two local spread-spectrum codes with a relative spacing of 1 chip time (i.e. the early and late code) or 0.1 chip for narrow correlator. Each chip of C/A code is about 977.5ns (1/1023MHz) long. The output of the two correlation processes are differenced to form a discriminator function known as S-Curve. Multipath signals tend to distort the shape of S-curve, which result in a tracking error. A more detailed discussion on the receiver correlation processes can be found in [12] and [13].

The magnitude of multipath error mainly depends on the amplitude, time delay, and phase of the multipath signal relative to the LOS signal. The steps taken to calculate the map of multipath error displayed in Fig.11 are as follows:

1. Given a particular satellite with an elevation angle, in this study, SV15 was chosen at an elevation angle of 15 degrees. Note that the visibility of this particular satellite is also shown in Fig.11.
2. In this study, the power of the reflected signal is set to 0.5 relative to the direct signal, as the reflection coefficient for concrete at the propagation angel of 15

degrees is about 0.5. A detailed discussion and derivation of the reflection coefficients for various materials can be found in [14].

3. Only one single reflected ray is considered when the direct signal is present. The reflected-only signals are not displayed in Fig.11.

4. The additional length the reflected signal travels compared to the direct signal is divided by the speed of light, the resultant quantity is the time delay. The phase delay is simply the additional length divided by 19cm which is the wavelength of GPS L1. Alternatively, the phase delay can be set to either 180 or 0 degrees in order to assess the absolute maximum multipath error.

5. Every cell centre in the 1m LiDAR data is used as an observation point (i.e. receiver position) at which ray tracing is performed in order to detect the reflected rays. As such, the resultant prediction map has a very high resolution of 1m.

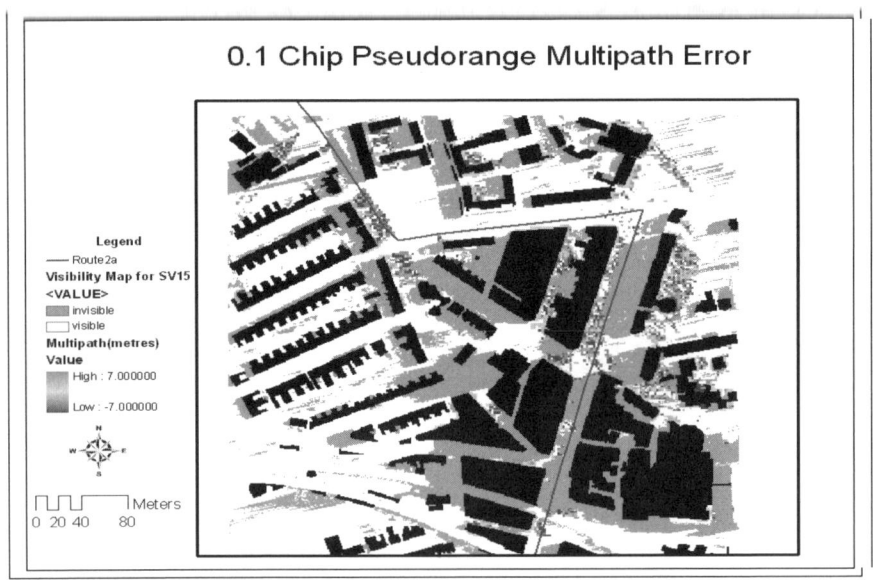

Fig. 11. 0.1 chip pseudorange multipath error in metres

As described in Fig.11, the reflections are more likely to occur when the receiver is placed close to a building. The pseudorange multipath map gives clear information of the possible occurrences of multipath due to one single reflection.

6 Conclusions

In conclusion, the use of LiDAR data and 2D building footprints can help model and predict GPS multipath effects. The methodology presented in this paper is based on the well-established theories in the domain of GIS, GPS and radiowave propagation. Furthermore, the terrain data in raster format used in this work is compatible to common GIS applications. Either a Triangulated Irregular Network (TIN) or regular grid can be used in conjunction with the building footprints. Therefore, the methodology does not require a specific data structure for 3D city models. Future work will focus on the detailed study of the noise characteristics in different terrain environments using the developed ray tracing model.

References

1. Taylor, G., Li. J., Kidner, D.B., Brunsdon, C. and Ware, J.M. : Modelling and Prediction of GPS Availability with Digital Photogrammetry and LiDAR, The International Journal of Geographical Information Science, (2006), (in press).
2. Vrhovski, D.: Satellite visibility in simulating urban satellite positioning-based road user charging, proceedings of ION GPS/GNSS 2003, Portland, Oregon, USA (2003).
3. Pratt,A,R.: Performance of Multi-path Mitigation Techniques at Low Signal to Noise Ratios, proceedings of ION GPS/GNSS 2004, Long Beach, California, USA (2004).
4. Strang, G.., and Borre, K.: Linear Algebra, Geodesy, and GPS (book), Wellesley-Cambridge Press, (1997).
5. Rao. B.R. Sarma A.D. and Kumar : Techniques to reduce multipath GPS signals, Current Science, Vol.90 (2), (2006).
6. Ordnance Survey, OS MasterMap user guide, Available in World Wide Web http://www.ordnancesurvey.co.uk/oswebsite/products/osmastermap/guides/userguide.html, accessed on 12 Jan 2006.
7. Brunner, F.K., Hartinger, H. and Troyer, L.: GPS signal diffraction modelling: the stochastic SIGMA-D model, Journal of Geodesy, Vol. 73, (1999), 259-267.
8. BÉTAILLE, D.: A testing methodology for GPS Phase mitigation techniques. Proceedings of ION-GPS, Portland, Oregon, Institute of Navigation,USA,(2003).
9. Taylor, G. and Blewitt, G.: Intelligent Positioning: GPS-GIS unification, Wiley (book), (2006).
10. Catedra, M.F. and Arriaga, J.P.: Cell Planning for wireless communications (book), Artech House Publishers, London, (1999).
11. Parsons, J. D.: The Mobile Radio Propagation Channel (book), Chichester, Wiley, (2000) pp36-37.
12. Braasch, M. S.: Multipath Effects. in Chapter 14 of Global Positioning System: Theory and Applications, Vol. 1, edited by B. Parkinson, J. Spilker, Jr., P. Axelrad and P. Enge, American Institute of Aeronautics and Astronautics, Washington, D.C (1996).
13. Van Nee, R.: Multipath Effects on GPS Code Phase Measurement, Proceedings of ION GPS-91, Albuquerque, Washington, DC, USA, (1991), pp. 915-924
14. Hannah B. M.: Modelling and Simulation of GPS multipath propagation, PhD thesis, Queensland University of Technology, (2001).

The Web Integration of the GPS+GPRS+GIS Tracking System and Real-Time Monitoring System Based on MAS

Ye Lei[1,2] and Lin Hui[3]

[1] Management College, Shanghai Business School,
2271 West Zhongshan Road, Shanghai 200235, P.R. China
yelei@sbs.edu.cn
[2] Shanghai GALILEO Industries Ltd.,
6F, Bldg. 33, 680, Guiping Road, Shangahi 200233, P.R. China
yel@shgalileo.com.cn
[3] Joint Laboratory for GeoInformation Science, the Chinese University of Hong Kong,
Rm. 615, Esther Lee Building, Chung Chi College, CUHK, N. T., Hong Kong
huilin@cuhk.edu.hk

Abstract. In this paper, the Multi Agent System (MAS) Architecture, GPS, GIS, and Wireless Communication technologies were discussed. New application architecture of complex vehicle Location Based Service (LBS), and Navigation and Intelligent Transportation systems based on the MAS architecture were proposed. The implementation and Web performance of this methodology on the urban garbage trucks management in the Shanghai Putuo district were introduced. The final system testing results were evaluated. And the future potentials of the MAS based approach to solve complex urban management problems and monitoring information network systems were prospected.

Keywords: Software Agent, Multi Agent Systems, GPS, GIS, Wireless Communication.

1 Introduction

In the research field of the Software Agent, implementations are more advanced than other theory studies. As an important computation and construction unit of the digital world (with the development of the computer and communication hardware and software), there have been many innovations and diversified applications about the Agent system, first from a single Agent, then Multi Agents, and now Mobile Agents.

Actually, Agent technology is becoming the most important component in constructing the socially organized system of the cyber world. Agent applications include information services, multi-dimension designs, robots, e-business, computer aided cooperation, computer games, education and training, intelligent environment, society simulation, artificial life and so on.

Location Based Service (LBS), Navigation and Intelligent Transportation are the hot research topics now. There are many wonderful solution methods emphasis on above topics, e.g. PDA (Personal Digital Assistant)/Pocket PC and GPS (Global

J.D. Carswell and T. Tezuka (Eds.): W2GIS 2006, LNCS 4295, pp. 54–65, 2006.

Positioning System) integration, embedded navigation software development, GIS(Geographic Information System) based transportation commanding center building, and so on. But our objective using Agent technology and Multi Agent System architecture is to find a total solution approach to cover the main questions in this area.

Research on Location Based Service (LBS) and Navigation and Intelligent Transportation generally tackle three basic questions (Shih-lung Shaw, 2005):

1. Where am I? (How to calculate my position)
2. What is around me? (How to express the objects and environment around some place)
3. How can I go to that place? (Best way of getting from A to B)

Question 1 can be solved now by the integration of the GPS (Global Positioning System) or Satellite Positioning System, sensor network positioning and embedded GIS (Geographic Information System) software. The position information can now be transferred to the users' PDA or smart mobile phone using Short Message Service (SMS) or push technologies (GPRS- General Packet Radio Service, WAP- Wireless Application Protocol, etc.).

Question 3 can be tackled by the research on network analysis models and algorithm for way-finding. Some familiar way-finding algorithms [1][2][3] include Depth-first searching algorithm based on the network limitations; Dynamic programming algorithm in an acyclic network graph with direction identification; Dijkstra algorithm based on the adjacency matrix; Maximum dependence edge algorithm; and Dijkstra algorithm based on the greed and heuristic game etc.

But for question 2, the transportation monitoring system distributed at different road junctions and crosses is a better method to offering the real-time environment video frequency information around us.

So new methodology should be proposed, which integrates GPS installation in vehicles, data transmission through wireless network services (e.g. SMS- Short Message Service, GPRS- General Packet Radio Service); direction promotions supported by the GIS based way-finding algorithms; and Web real-time video capture systems. This methodology gives us the possibility to solve total Location Based Service (LBS), Navigation and Intelligent Transportation questions.

This paper proposes a MAS architecture implementation to solve these two important problems in the Web integration of the GPS+GIS+GPRS Tracking System and Real-time Monitoring System.

The Agent here is not only a modern, advanced, computation technology as people wish, but also a new solution methodology and a new concept model and implementation tool to study the complex, distribution and interaction system.

2 Software Agent and Multi Agent Systems

Referenced to the Agent concept by M. Minsky (1994), the famous computer scientist and the one of the Artificial Intelligence founders, Software Agent are self governed software/integrated software package with special skills, with regard to one computer system. When you need accomplish a task without any knowledge about the inside process and the software, viz. the software/ integrated software package running as a black box, it could be defined a Software Agent.

In other research works on the characteristics of Agents, the most popular and classical theory is the discussion about the Agent's "weak definition" and "strong definition" (Wooldridge 1995, 1997; Nwana 1996). After analyzing some typical research re-ports and application systems, based on the description and definition of Agents, the basic characters of Agents were found. These characteristics include Interactivity, be task/goal driven, be autonomous & controllable, and reactive (Liu Da-you 2000; Sun Yu-bin 2000). These four characteristics are the basic Agent characters, but they could have other characters according to the application situation, such as mobility, veracity, self-adaptability, Communicative, sensitive, self-rebooting, Self-benefiting etc. [4] [5] [6] [7]

In the practical research and program of Software Agents, all the characters are not necessarily built into one Agent System or Multi Agent System. Usually we choose some characters of Agents to build the system according to the practical application.

But the reactive, interactive, task/goal driven, autonomous & controllable must be regarded as the basic technology and theory of Agents.

Any Software Agent is designed for the user's special task and goal, and the running mechanism of the Agent is based on the task/goal driven characteristic, so how to build the Task/Goal Driver is the key part of the Software Agent. The most popular characteristics of Task/Goal Driver include Data Driven and Message Driven. The Data Driven method is usually used in Management Information Systems, because all the operations and actions are activated by data. The Message Driven characteristic is often used in the Windows Operating System, and the functions or actions are based on watching and triggering. The Software Agent manages the goal of the system as a whole, so it uses the Task/Goal Driver based in the Data Driven, Message Driven and other methods (the relationship see also Fig. 1).

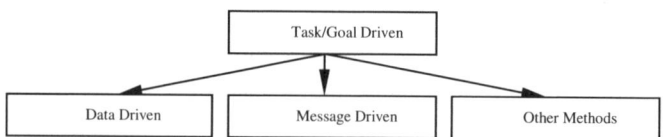

Fig. 1. Task/Goal Driven buildup [8]

In other words, a Software Agent is a kind of program or computing entities, which can sense the environment, self-run, realize the task/goal given by its designer or user, and give the right reaction.

2.1 The Common Agent Architecture Design

As an important computation unit and application component of the digital world, Agent creation should be described as a practical engineering technology. Usually, the architecture of Agents are always inter-infiltrated with software engineering main technologies, such as Object Oriented Software engineering (OOSE). Such technologies can implement special functions according to the Agent concept, such as

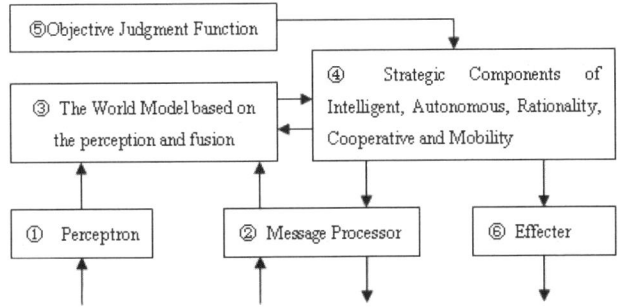

Fig. 2. The Common Agent Architecture [9] [10] [11]

autonomy, rationality, co-operation and mobility. The six parts of Agent architecture are summarized in figure 3.

The Perceptron monitors the environment and system status, and produces the Agent input information.

The message processor is responsible for the information exchange with other Agents. This message interactive mechanism is based on the bit groupings and some self-expression texts. Usually TCP/IP (Transmission Control Protocol and Internet Protocol) apply to describe the message processor, especially in Web service. The contents of the socket interface include service thread, message thread, custom thread, timer, receive buffer, send buffer, send schedule and receive schedule etc.

In a constant period, this model acquired information from the Perceptron and Message Processor, and fuses the multi data and information to some congruous environment status.

The Strategic Components of Agents are the function groups including the prop steps they use to handle problems (or to decompose one problem to several sub-questions). Some functions are expressing the location of Agents and calculating the range of model variable values. The objective judgment function is built variously in order to solve different application problems.

Two values compared actualize a simple judgment function (e.g. comparing the GPS signal with GIS objects coordinates in order to judge the location). There are many judgment methods that may include fuzzy set, range evaluation, multi attributes assessment and constraint condition judgment etc.

The Effecter is the output of system. Information produced by an Agent could affect the outside object and the Agent itself. One Agent system contains many executors in order to accomplish the objective. The computation results of Agent components are inner variables, and will be converted to executable programs and arrange the execution sequences. In the software system, the Effecter might be a command word, a network packet, an alarm E-mail or some audible or visible alarm.

2.2 Multi Agent Systems

Like many developing technologies, applications and implementations keep ahead of the theoretic progresses about Software Agents. As one important computation and organization unit in the digital world, Agent systems produce many creative applications from single Software Agent systems to Multi Agent cooperating systems, and the Mobile Agent applications. Agent technology is becoming the socially organized system and component of the digital world, based on the intercross and integration of Artificial Intelligence (AI), Object Oriented (OO) and Distributed Computation Network (DCN). The applications of MAS cover Information Services, Multi Dimension Interface Designs, Robots, E-Commerce, Computer Aided Cooperation, Computer Games, Education and Training, and Simulating Society and Artificial Life etc. These applications not only give us an advanced computation technology, but also show a brand-new thinking method to solve the complex system problems [12] [13].Based on the MAS architecture, a new complex system concept model was proposed, and many developers are doing their creative research in this field. The Belief-Desire-Intention Theory (proposed by Bratman) is considered an important fundamental theory.

The architecture of MAS (see Figure 4) describes the basic components, the functions, the relationships and the communication mechanisms between different Agents. It also shows the information interaction, the control relationship, the network surrounding, and the Agents abilities to solve problems.

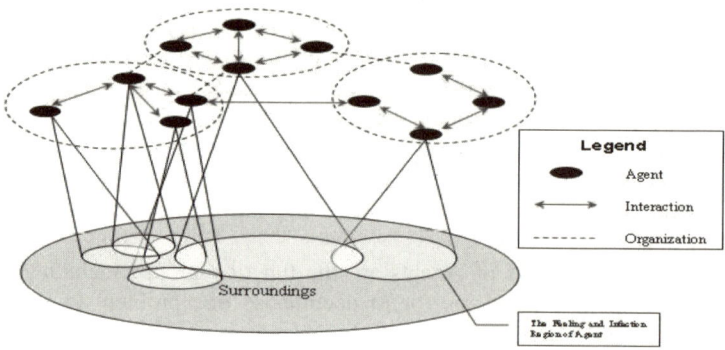

Fig. 3. The Common Agent Architecture [14]

Usually the MAS architectures are grouped with 3 types, Deliberative Architecture, Reactive Architecture and Hybrid Architecture.

2.3 MAS Based Vehicle Management System Architecture

As we discussed in the introduction, the system integration objective is contained in GPS systems and installed in different mobile vehicles. The GIS system offers the static environment information and location coordinates, using the GPRS system as the communication channel. It provides for the real-time video monitoring systems that constantly acquire and publish the dynamic traffic information. The information is made up of the settled road cross, and all of the four different systems must be developed and integrated in the Browse/ Server architecture in order to offer anyone, anytime, anywhere information Web service.

Because each vehicle might move anywhere with any purpose, object mobility must be supported. And as a complex integration software system is involved in

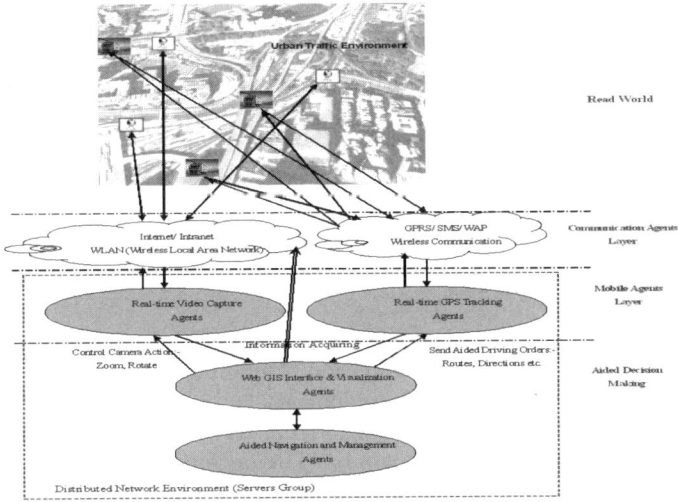

Fig. 4. Vehicle Management System Based on Contract Net MAS Architecture

different sciences and technologies, Agents and Multi Agent Systems are selected to meet the architecture designs (see Figure 5).

The communication Agents layer contains two different communication archi-tectures. One is a TCP/IP based Internet/Intranet/WLAN communication network, which focuses on the Web Service Oriented technologies based on XML. The other is the commercial wireless communication network (e.g. GSM- Global System for Mobile communications, CDMA- Code Division Multiple Access, SMS- Short Message Service, 3G- third generation etc.), which focuses on information transmission package technologies including TCP/IP socket, WAP push, database middleware etc.

The mobile Agents layer contains real-time video capture Agents and GPS Tracking Agents. Media Flow process and publication technologies should be the key to the real-time video capture Agents. GPS related technologies support the GPS tracking Agents. Web based information service technologies support the communication between Agents, which include database middleware, TCP/IP socket, Web Service etc.

The aided decision-making layer contains the Web GIS interface, visualization Agents; and Aided Navigation and Management Agents. The Web GIS technologies support the creation of the first Agents group. The way-finding algorithms and cooperation technologies with work flow system and GIS support the aided Navigation and Management Agents creation.

Except the communication between the Web GIS Interface and the Visualization Agents to the Communication layers, the communication channels between different components and Agents are two-way. The reason for the one-way communication channel is that the status of the communication layers are never shown on the Web GIS interface.

This architecture is classified into Contract Net Hybrid MAS Architecture. The Web GIS Agent is the administrator of the lower level Agents including communication Agents, video capture reactive Agents and GPS tracking mobile Agents. At the same time, the Web GIS Agent also takes responsibility for executing the order coming from the aided navigation and management Agents.

3 Other Key Technologies

3.1 Vehicle GPS Technologies

Vehicle GPS is an integration of software and hardware systems installed on different vehicle with GPS navigation, positioning, and monitoring of the status of driving and controlling a vehicle and its electronic equipment.

One set of Vehicle GPS system contains the GPS satellite signal receiver and processing all-in-one machine, a Laptop or PDA or Special Vehicle PC, Electronic Maps, the Bracket, and the Vehicle Power Inverter or Charger provides the power for the whole system.

3.2 Communication Technologies

3.2.1 Socket in TCP/IP Protocol

TCP/IP is a set of complete protocols supported by multi-operating systems, which are used in the Windows series, UNIX, Linux etc. Different types of computers can do reliable data exchange following TCP/IP in the network. Sockets are located on the transportation layer in the 4-layer model structure of TCP/IP. Sockets supply the uniform interface with different application software (the relationships see also Fig. 10), such as Telnet, FTP, Ping, HTTP, WWW etc.

Sockets are also a software entity. They supported IPC (in-process communication by UNIX) of the distributed environment, and offer the basic component for IPC. Windows was used to do the test, so the socket software was WinSock interface. In many advanced computer languages, WinSock is capsulated in the bottom layer of API by components. Sockets accomplish the network communication according to the C/S mode.

3.2.2 XML

There have been multitudinous benchmarks for evaluating the DBMS performance in various application areas. On one hand, XML becomes a standard for data interchange [15], penetrating virtually all areas of Internet applications, and bringing about massive amounts of data. XML allows one to specify the content and structure of a document in a way that lets one generate particular presentations as needed [16].

The situation is that XML-based solutions are becoming the preferred choice for most GIS vendors and users [17]. There is an increasing need for the management and exchange of spatial data in modern applications with the emergence of XML as a standard for information interchange on the Web. Moreover, Geographic Extensible Markup Language (GML, the XML application in the specialized domain) also becomes a standard for geospatial data sharing and transport over the web [18].

3.2.3 GPRS [19]

General packet radio service (GPRS) is a packet-based wireless data communication service designed to replace the current circuit-switched services available. These services are usually available on the second-generation global system for mobile communications (GSM) and time division multiple access (TDMA) IS-136 networks. GSM and TDMA networks are designed for voice communication, dividing the available bandwidth into multiple channels, each of which is constantly allocated to an individual call (circuit-switched). These channels can be used for the purpose of data transmission, but they only provide a maximum transmission speed of around 9.6Kbps (kilobits per second).

As a packet-switched technology, GPRS supports the internet protocol (IP) and X.25 (packet-switched standards currently used in wire line communications). As such, any service that is used on the fixed internet today will also be able to be used over GPRS. Because GPRS uses the same protocols as the internet, the networks can be seen as subsets of the internet, with the GPRS devices as hosts, potentially with their own IP addresses.

Enabling GPRS on a GSM or TDMA network requires the addition of two core modules, the Gateway GPRS Service Node (GGSN) and the Serving GPRS Service Node (SGSN). The GGSN acts as a gateway between the GPRS network and the public data networks such as IP and X.25. They also connect to other GPRS networks to enable roaming. The SGSN provides packet routing to all of the users in its service area.

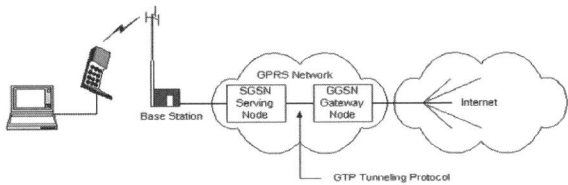

Fig. 5. GPRS configuration diagram

As well as the addition of these nodes, GSM and TDMA networks have to have several extra upgrades to cope with GPRS traffic. Packet control units have to be added and mobility management, air interface and security upgrades have to be performed.

3.2.4 Middleware and Database Middleware [20]

Middleware (software that functions as a translation layer) sits between an application residing on one server and any number of clients that want access to that application. In short, middleware allows users to interact with one another and with applications in a heterogeneous computing environment.

The types of middleware include database middleware, application server middleware, message-oriented middleware, transaction-processing monitors and Web middleware.

Database middleware only enables applications to communicate with one or more local or remote databases. It doesn't transfer calls or objects. And while database-oriented middleware is easy to deploy and relatively inexpensive, it doesn't include features found in more complex software products.

Database middleware doesn't allow for two-way communication between servers and clients. Servers can't initiate contact with clients, they can only respond when asked.

Application server middleware is a Web-based application server that provides interfaces to a wide variety of applications and is used as middleware between browser and legacy systems.

Messaging-oriented middleware provides an interface between client and server applications, allowing them to send data back and forth intermittently.

Messaging middleware is similar to an e-mail system, except that it sends data between applications. If the target computer isn't available, the middleware stores the data in a message queue until the machine becomes available.

3.3 GIS and Web GIS

The geographic information system (GIS) is a technical system of geography information science, using the theories and methods of system engineering and information sciences to acquire, store, manage, analyze, distribute and utilize spatial information with the support of computer software and hardware. GIS is widely used and has become a powerful tool in many areas, such as urban planning, city infrastructure management, traffic control etc.

Web GIS is the mainstream development and application of GIS technology. Web GIS is used as the provider of spatial data browse, query and analysis. The application structures of Web GIS mainly contain C/S and B/S architectures. When using the C/S architecture, part of the application software must be deployed in the client side, and the request of clients must be completed by the cooperation of the client and server. But for the B/S structure, the requests of users are processed on the server side, and there is no need to deploy system software on the browser side.

4 An Implementation Sample

In order to support the architecture and technologies introduced in this paper, a unified command system was built in Putuo district of Shanghai, P. R. China.

The objective functions of this integrated system contained GPS+GIS+GPRS garbage truck real-time tracking and commanding, key refuse collection fields and main road cross real-time video monitoring and operational information automation. Through the uniformed Web interface, anyone could query the urban environmental information at anytime, anywhere. In addition, remote sensing imaging from the Quick-Bird satellite every half year is used as the geo-referenced background (see Figure 7).

The Special Functions of this integrated system (besides the common Web GIS functions) include:

- Selection cooperation with real-time video-monitoring systems;
- RS imaging and GIS attributed integration multi-scale visualization;
- GPS tracking and monitoring in the B/S framework (see Figure 8);
- And Warning about filled cesspools

Fig. 6. The Screen Copy of Real-time Monitoring of Mobile Vehicles and Video Capture of a Road Cross

Fig. 7. Mobile Vehicle Historic GPS Track

The system Software:

- Microsoft Windows Server 2003/ Windows XP professional
- Microsoft SQL Server 2000 personal
- J2SDK 1.4.2
- IBM WebSphere/ Apache Tomcat 5.x
- ESRI ArcIMS 9.x

A Servers Group with 6 Dell 2850i was also built, one server machine was used for Communication Agents, one for real-time video capture Agents, one for GPS Agents, one for Web GIS Agents, one for the non-spatial information database, one for the spatial database (including remote sensing images), and one for system backup.

After the system testing (testing result see Table 1) and commissioning at the beginning of 2005, this system is now offering its service of supporting about 120 garbage trucks tracking daily in Shanghai. And a demonstration simulation system was also built in JLGIS of CUHK, Hong Kong in March 2005.

Table 1. Table captions should be placed above the table

		Merit	Testing Value	Explanation
Functionality	Compatibility	Sufficiency	0.978	1- Best
		Integrity	0.933	
		Coverage Factor	0.911	
	Exactitude	Evaluation	0.167	0- Best
Reliability	Maturity	Fault density	0.0476	
		Fault Settlement	1	
		Testing coverage factor	1	
	Recovery	Easy to Reboot	1	1- Best
Easy to use	Easy to Understand	Integrality of Description	1	
		Comprehensive Function	1	
	Easy to Operate	Intelligibility of System Messages	0	0- Best
Efficiency	Time Response	Average values	12.85s	
Transferability	Adaptability	Hardware	1	1- Best
		System Software	1	

The testing results was coming from one-week uninterruptedly supervision of this system, and it showed enough evidence to demonstrate the superiority of MAS based complex software system integration. The work reported in this article is also one of the most important parts of the 2010 World Expos project (05DZ05808) funded by the Shanghai Science and Technology Development Funds.

5 Conclusions

Usually the researchers in Location Based Service (LBS), Navigation and Intelligent Transportation area study the different approaches around the three key questions mentioned at the beginning of this paper separately. But Agent and Multi Agent System architectures, a new methodology to solve the complex system problems, have proved that the Location Based Service (LBS), and Navigation and Intelligent Transportation questions should be treated as a complete system.

The potentials of those key technologies and the integrated architecture discussed in this paper have not been explored fully. They have only been tested through the example of garbage truck real-time tracking and commanding. Continuous research is needed to further enhance the operability of the methodology proposed in this paper, such as solving the problems in transportation navigation, urban emergency reactions, and information systems integrated with real-time control systems in industry areas and so on.

References

1. Stig Nordbeck, Bengt Rytedt. Computer Cartograohy Shortest route Programs, Sweden: The Royal University of Lund, 1969.
2. Rune. The A* algorithm: Comparison to other common path-finders, http://www.cs.auc. dk/~rune/ FE1101/litt/A star/PathFinders.html, 1997.
3. Dijkstra E W. A note on two problems in connexion with graphs. Numerische Mathmatik, 1959, (1),269-271
4. M Wooldridge, N R Jennings. Intelligent agents: Theory and Practice. Knowledge Engineering Review, 1995, 10(2): 115-152.
5. M Wooldridge, Agent-Based software engneering. IEEE Transactions on Software Engineering, 1997, 144(1): 26-37.
6. M Wooldridge, N R Jennings, D Kinny. A methodology for agent-oriented analysis and design. http://www.cosm.ecs.soton.ac.uk/nrj/pubs.html
7. Nwana H. Software Agent: an overview. Knowledge Engineering Review, 1996, 11(3): 205-244.
8. Sun Yubin, Lin Zuoquan. Software Agent. Computing Technology and Automation, 2000, 19(1): 75-79.
9. Gasser L. Agents and Concurrent Objects. IEEE Concurrency, 1998, 6(4): 74-77.
10. Poggi A.DAISY: An Object Oriented System for Distributed Artificial Intelligence. Wooldridge M, Jennings N R, eds. Intelligent Agents: Theories, Architectures, and Languages (890) [C]. Springer Verlag: Heidelberg, Germany, 1995: 341-354.
11. FENG Shan , TANG Chao , MIN Jun . Software agent Software engineering approach Theoretic model Problem solving. Systems Engineering and Electronics, Vol. 24, No.12 2002: 96-99.
12. M Tokoro. Computational Field Model: Toward a New Computing Model/Methodology for Open Distributed Environment. Proceeding of 2nd IEEE Workshop on Future Trends in Distributed Computing System, Sept 1990.
13. E Osawa. A Scheme for Agent Collaboration in Open Multi-Agent Environment. Proceeding of IJCAI's 93, August 1993: 352-358.
14. He Yanxiang, Chen Xinmeng, The Design and application of Agent and Multi Agent System, Wuhan University Press, June 2001,   ISBN 7-307-03163-9/TP· 101: P1-21.
15. T. Bray, J. Paoli, C.M. Sperberg-McQueen, and E. Maler, Extensible Markup Language 1.0 (2nd edition), 2000.
16. http://www.unicode.org/iuc/iuc13/k2/sld01004.htm
17. Z.R. Peng and C.R. Zhang, The Roles of GML, SVG, and WFS Specifications in the Development of Internet GIS Journal of Geographical Systems, Vol. 6, No. 2: 95 - 116.
18. S. Cox, P. Daisey, R. Lake, C. Portele, and A. Whiteside, OpenGIS® Geography Markup Language Implementation Specification, version 3.1.0, OGC, Inc., Feb, 2004.
19. http://www.mobilecomms-technology.com/projects/gprs/
20. Linda Rosencrance, Middleware, http://www.computerworld.com/softwaretopics/software/appdev/story/0,10801,52066,00.html

A Management System of Street Trees by Using RFID

Eui-myoung Kim[1], Mu-wook Pyeon[2], Min-Soo Kang[1], and Jae-sun Park[2]

[1] Korea geoSpatial Information & Communication Co. Ltd, Seoul, Korea
{kemyoung, mskhang}@ksic.net
[2] Department of Civil Engineering, Konkuk University, Seoul, Korea
{neptune, xteen88}@konkuk.ac.kr

Abstract. To manage cities by using Ubiquitous technology, a U-city has recently been gaining attention in Korea. With a great deal of help from the Central Government and local governments, a series of trials to integrate the traditional complex facilities and natural items with the technical elements of the ubiquitous environment such as RFID, SoC and USN are actively in progress. Particularly, in the case of combining with the existing GIS system, they will be very efficient. and In this study, as an actual example of the application of RFID, wireless telecommunications, and GIS, RFID is introduced to manage street trees in cities and UFID is studied for a new address system in Korea. To enhance the existing management system, GPS, CDMA and a web information system are constructed and finally the three dimensional GIS system is developed.

Keywords: RFID, Ubiquitous, Facility Management, U-city, UIS, UFID.

1 Introduction

The Korean government recently selected Ubiquitous technology as the main item of the top-10 growth engine industries to lead Korea. As a result, each industry is positively seeking the introduction of ubiquitous technology to its environment. In particular, the construction of a U-City that enhances the quality of life and ensures the effective management of cities through making the structures of a city intelligent is in its initial state.

The urbanization of Korea is up to 80.8% as of 2005. Particularly, cities in Korea are densely populated, which naturally allows the Central Government and local governments to operate GIS-based UIS (Urban Information System) to manage cities efficiently. Currently the UIS is operated successfully and used for the management of main infrastructures like roads, buildings, water supplies and sewage system, and land. It is likely that this GIS-based management system will become a strong basis for cities of Korea to enter the ubiquitous environment in the near future.

In this study, as an actual example of the application of RFID, wireless telecommunication technology, and GIS technology for the management of facilities in cities, RFID for managing street trees and a new address system suited to the Korean situation were designed and introduced. In addition, to improve the existing system of street trees, GPS, CDMA, and a Web Information System are constructed. Finally, a three-dimensional GIS system is developed.

J.D. Carswell and T. Tezuka (Eds.): W2GIS 2006, LNCS 4295, pp. 66–75, 2006.
© Springer-Verlag Berlin Heidelberg 2006

Until now, the existing system has been mainly operated by on-site works, where stainless steel labels are used. In this study, however, a U-street trees system is proposed for the systematic and efficient management of street trees by using advanced GIS(Geographic Information System) and up-to-date information telecommunication technologies.

2 Design of the Management System of Street Trees

2.1 Functional Requirements Analysis

This is the step for understanding the demands from users to realize an optimized system. The main points of this step for the realization of the management system of street trees are shown in Table. 1.

Table 1. The main points of the Functional Requirements analysis step

Main points	Details
To apply the real-time update function	. Necessity of the application of wireless telecommunication technology
To include car control function	. Necessity of GPS device
To include 3D modeling function	. Necessity of a three-dimensional model of trees lining street . Necessity of software for three-dimensional GIS function
Hardware specification definition	. Determines frequency by requesting RFID recognition distance . Determines hardware like (PDA: Personal Digital Assistant)
How to mount RFID	. Maintains a consistency in case of on-sites work

2.2 System Design and Development

The management system of street trees is generally divided into a PDA system, for the on-site management of street trees, and a two-dimensional (or three-dimensional) web server development for workers responsible for managing street trees.

In the PDA system, the locations and shapes of on-site street trees are expressed on an electronic map. Searching, amending and inputting the information of street trees in charge could be done on a terminal by using mobile computing technology. The amended contents are input in real-time through a management server.

The web information system could control cars on-sites and amend and add the existing information. In car controlling, by using a PDA having interoperability with GPS, the locations and status of cars, for example, spraying, moving or trimming, could be identified in real-time and all the contents recorded in a database. The web information system could manage information remotely with an interactive system that has services of search and inquiry regarding the information of street trees.

2.3 Building of a Street Trees DB

The construction of a DB is based on the data of the Large Tree-Registration System (trees with more than 20cm diameter should be registered), operated by Seoul. A field study should be done to identify whether or not main data and data on-site are the same and should add field and input data to realize the system.

The main contents of the database to be managed by the management system of street trees are as follows.

- Environment Information: Management Number, Street Name, Group Name, Administrative District, RFID Attachment, etc.
- Tree Information: Tree Kinds, Size, State, etc. [1]
- Management Information: Planting Date, Safe Cover, Support, Lower Vegetation, Soil Management, Pest Control, Herbicide, Water and Fertilizer, etc,
- Code Information: Street Trees, Cars and Species of Trees

3 UFID (Unique Feature IDentifier) Design for the System

A UFID (Unique Feature Identifier) is a unique identifier, which is under development for the purpose of managing national land with a single unification system by the Ministry of Construction and Transportation. UFID gives numbers to the land space and identifies the locations and contents of any objects that are wanted with an identifier.[2] Therefore, UFID not only integrates sensors for real-time management, telecommunications technology, altitude information along with location information, and the information from management organizations but also connects with GIS, which could systematically manage and apply underground facilities, underground space and natural geographical items, and artificial facilities on the ground.

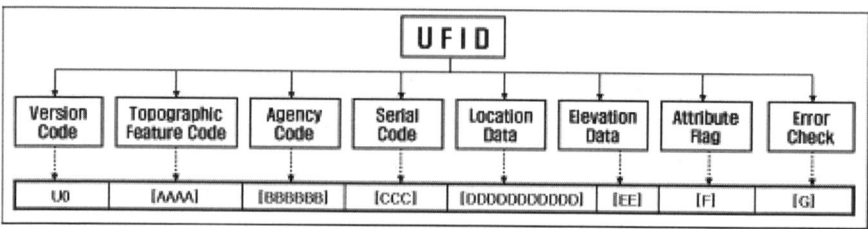

Fig. 1. Structure of UFID

The UFID is used as an identifier necessary for LBS and a ubiquitous environment. UFID is based on the UFID system from the National Basic Geographical Information and is suggested by applying a serial number field and management organizations. (Fig. 1)

1) Version Code
According to the National Basic Geographical Information UFID System, a two-digit number is used through the combination of the version number of UFID and its U.

2) Topographical Feature Code

According to the numerical map ver2.0 system [3], one alphabet character and small category number 3 are expressed. Street trees have no corresponding classification system but are expressed as D003 code.

Table 2. Numerical map ver2.0 geographical things classification system

Large classification name	Road	Building	Facility	Gardening	Water system	Topography
Large classification code	A	B	C	D	E	F
Small classification number	22	2	55	4	8	5

3) Agency Code

The construction of a management organization field is divided into three types

(1) Management Subject
   - Each city hall, district office and other organization for managing street trees
   - How granted: Organization Integration Code of the Ministry of Government Administration and Home Affairs (7 digit number)
(2) Administrative district
   - Administrative district where street trees are located
   - How granted: Administrative-dong code (Metropolitan-si/do<2> + si/gun/gu<3> + eup/myeon/dong<3>)
(3) Road Management Number (selected)
   - Road number where street trees are located
   (Metropolitan/do<2> + si/gun/gu<3> + serial number<5>)

In this case, the road management numbers make it easy to understand the locations and distributions of street trees.

Through inputting the road management number into the UFID, it could be connected with road names and a building number management system, which is very effective in

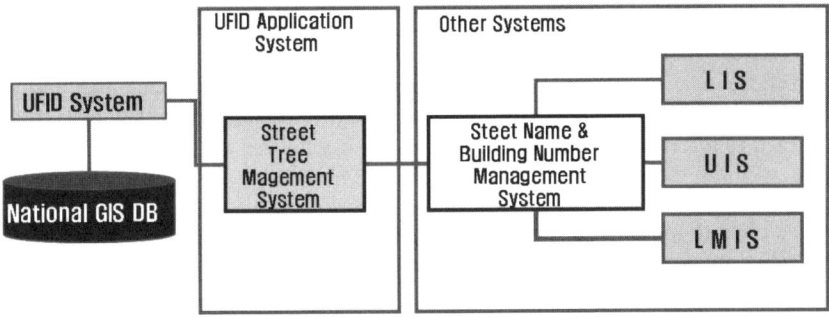

Fig. 2. Linkage of UFID and various systems

the case of linkage of other systems, because the Urban Information System (UIS) is also managed by the Korea Land Information System (KLIS), the Land Management Information System (LMIS) and local governments, based on roads. (Fig.2).

In case of adopting the road management numbers, however, there is no unity or system in the existing national road management number, so the road management number must be re-constructed as a single standard for UFID application. In addition, merely the adoption of a road management number cannot give the information of who controls the street trees and the administrative-dong where street trees are placed, but this problem could be solved by inputting the administrative-dong code and management subject code into the database of street trees.

4) Serial Code

A serial number field is combined with a management number and is composed of information that recognizes facilities. Street trees are generally lined along roads so characters and serial numbers that recognize the direction of Left (L), Middle (M) and Right (R) of roads are expressed.

The roads and building number management system DB of other regions confirm that the longest road in Korea is Bongsanjung2-gil in Daejeon City (117,356.82m) [4].

From this, by predicting the increase or decrease of the volume of street trees including the extension and establishment of roads, serial numbers express five-digit numbers (99999cases) * 3 (direction), that is, a total of 299997 numbers of street trees per road could be identified. Such a serial number system could be applied to all the management of street trees across the nation.

5) Location Data

Location Information consists of latitude and longitude. The longitude of Korea is 124°- 131°, constituting a total of 11 digit numbers, excluding the first two numbers. For example, Seoul has a latitude of 37°34′ 51″ and longitude of 126°59′34″; the result is 37345165934.

6) Elevation Data

The elevation information is not necessary for the management of street trees but is given an empty code 00 in consideration of the interoperability and extensibility of the future UFID.

7) Attribute Flag

In the case where specific additional attributes are required, the responsible organizations ask the organizations managing the UFID to amend and apply by attaching the specific contents. If there is no specific matter, expressed as 0 or as 1, interpreted as referring to the specific matter of the database.

8) Error Check

This is a code to identify the integrity of the UFID, system transmission and possible errors when manual operated. A checksum that applies the denary scale, storing the results, is how errors are identified.

The serial number and geographical items code are characters, so they are used by converting the characters into the decimal system.

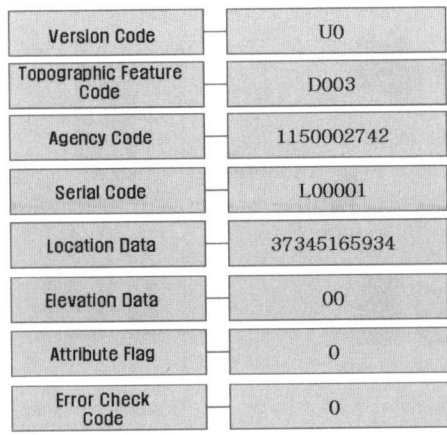

Version Code	U0
Topographic Feature Code	D003
Agency Code	1150002742
Serial Code	L00001
Location Data	37345165934
Elevation Data	00
Attribute Flag	0
Error Check Code	0

Fig. 3. Example of UFID Codes for a street tree

4 Development of the Ubiquitous-Based Management System of Street Trees

4.1 RFID Insertion and Data Input

This step is to insert the RFID into each one of the street trees and to input data into the RFID. The order of the process is chosen from the two types: first, inserting the RFID already containing data into street trees and second, inputting data after inserting the RFID into street trees.

The RFID uses 125KHz, 13.56MHz and 900MHz according to the recognized distance and in this study, 125KHz was selected[5]. In the case of 125KHz, the distance that the RFID could recognize is short but the size of the RFID's is small [6], which makes it possible to directly insert the RFID into very small holes in street trees, so that pedestrians cannot recognize it. The RFID is put on the existing stainless steel labels for easy recognition, considering the growth of trees.

4.2 System Integration

The integration of the up-to-date management system of street trees developed from the study that is shown in Fig. 4. The system is comprised of a recognition system that recognizes each street tree by using the RFID, a PDA mounted with a RFID reader that manages lots of information of street trees on-site, wireless telecommunication that transmits the information collected on-site and a web server for the inquiry and search of a variety of information on street trees.

For the management of street trees, first, each ID of street trees is recognized on-site by using a wireless recognition technology. At this time, cars responsible for controlling street trees use GPS and GIS and express the locations of workers and street trees on the electronic map of the PDA. The information of the locations, kinds

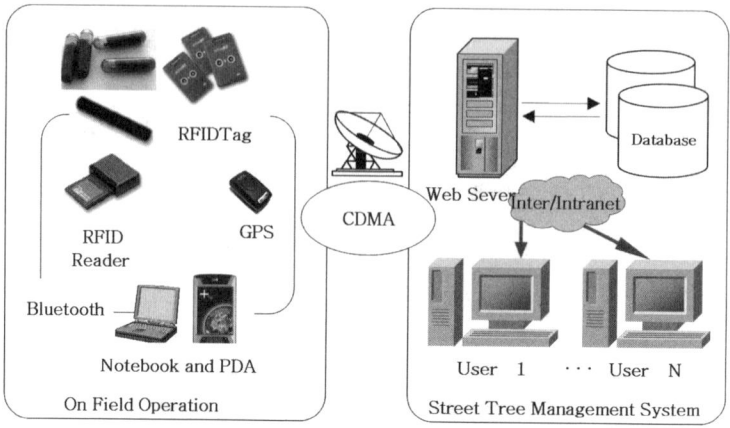

Fig. 4. The system integration (Hardware)

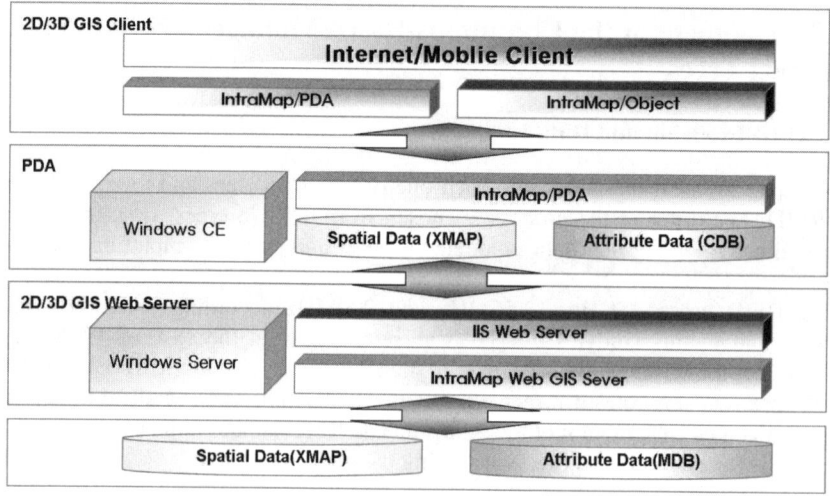

Fig. 5. The software architecture

and harmful insect attacks of each street tree is transmitted to the web server through the wireless telecommunication network and then those who manage the trees can search and manage on-line in their offices.

The management of street trees requires the information of the environment, the trees, management and codes through the on-site management system and web information system in general. Fig. 5 shows software architecture of the system. The existing management system could only separately recognized street trees of the initial state but the system proposed in this study makes possible a real-time management of street trees information on-site, as well as car control, information sharing and a three-dimensional model.

4.3 On-Site Management System of Street Trees

This system recognizes each tree and manages its information by using a PDA and RFID reader. Amended or updated information on street trees is transmitted through wireless telecommunication. This on-site system provides many functions for workers to perform recognition of street trees, information management, and information transmission on-site. (Fig.6).

Fig. 6. On-site management system of street trees (PDA)

(a) Recognition of street trees   (b) Information of street trees

Fig. 7. On-site management system of street trees (PDA)

(a) of Fig. 7 indicates separate trees chosen by a user in the two-dimensional map (b) shows the information about the chosen street trees.

Each tree could be directly selected by users on the map but could be instead automatically recognized through a RFID reader around RFID-mounted street trees and the information of street trees and management history could also be identified and amended.

The information has 25 items regarding management code, street code, group code, tree code, West and East, North and South, planting date, attachment date, status of trees, administration name, tree height, crown, DBH, measurement date, supporter, supporting type, supporting qualities, safe shell, frame type, frame qualities, plate type, plate colors, stain net, tree label and lower vegetation[7]. When selecting 'Move' at the tree information, the relevant map of street trees is extended. The management history indicates items that do necessary treatment for street tree management. If information of trees on-site changes, the changed information is transmitted to a control center through wireless telecommunication.

4.4 Web Information System

The Web Information System largely consists of the second and third-dimensions. Each street tree on the web information system could be selected by users or searched by a Search Window. Searching and amending street tree information has four types of street name, trees, planting date and administrative districts.

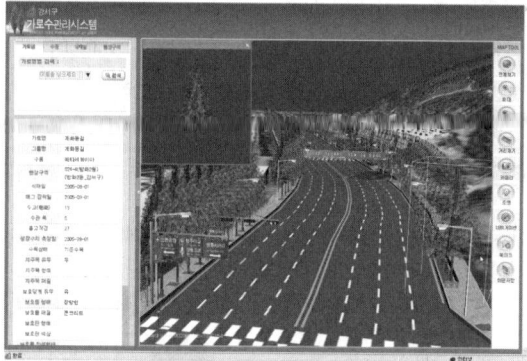

Fig. 8. Web-based road tree GIS system

Fig. 8 shows the search results and details of street trees selected through the name of 'Gaehwa-dong gil'. The three-dimensional web information system shows each street as a three-dimensional model of street trees so that planted trees could be identified at offices without directly visiting on-site. Those who process civil appeals could easily identify on-site information through the street tree management number through viewing the three-dimensional screen and managers also could treat them conveniently.

5 Conclusion

In this study, as an example of managing city facilities by applying ubiquitous and GIS technologies, a U-management system of street trees is developed. For this, RFID, wireless telecommunication and GIS technologies are introduced and a UFID suited to the new address system of Korea is designed. Further, to improve the existing management system of dual on-site and indoor works, GPS, CDMA and Web Information Systems are built up and finally, a three-dimensional GIS system is developed.

The purpose of this study lies in suggesting a case that the application of RFID, UFID and GIS could make an effective management of city facilities possible under a u-city environment in terms of the technical side of the issue[8]. It is expected that such performance will be comprehensively applied to the management of city facilities like streetlights, manholes, transportation facilities, and others.

In the near future, it will be necessary to study comprehensively the connection of the various city-composing elements of the UIS, GIS-based city management system and RFID technology. Particularly, studies of how to connect them with the highly advanced technologies like SoC(System on Chips) and USN(Ubiquitous Sensor Network) are required as well.

References

1. Gyeonggi Research Institute : Measures for the improvement of managing and planting trees in Gyeonggi-do, (2002) 12-15
2. Geographical Information System Institute ; Development of Application of (UFID) to Geographical items, (2002) 30-35
3. National Geographic Information Institute : Geographical Information Standardization-based Research-Study on Numerical Map Integration Standardization, Unification of geographical things code, (2002) 50-55
4. Jeong Seong-gwan, Park Gyoung-hun, Park Jin-soo, Kim Hee-nyun : 2000. Study of the application of code for the comprehensive management system of trees. Korea Association of Geographic Information Studies 3(1), (2000) 57-68
5. Park Gi-hwan : Ubiquitous RFID. Seongandang publication, (2005), 100-101
6. Lee Jae-yeol, Kim Seong-won, Choi Sang-young : Study of the application of RFID to military forces. MORSK 31(1), (2005) 58-72.
7. Sim Gyeong-gu, Her Sang-hyun : Study of the management of trees-for public officials responsible for managing trees lining streets. Korea Institute of Traditional Landscape Architecture 18(2), (2000) 81-88.
8. M.W. Pyeon, D.H. Ha, B.K. Lee, J.N. Lee : Technical Components for Ubiquitous Computing on Construction Process, Proceedings of PRIMA2005. (2005) 123-125

A Modular Neural Network Approach to Improve Map-Matched GPS Positioning

Marylin Winter and George Taylor

Faculty of Advanced Technology, University of Glamorgan, Pontypridd, Wales,
CF37 1DL, UK
mwinter@glam.ac.uk, getaylor@glam.ac.uk

Abstract. This paper provides an overview of work undertaken over the past two years to develop Artificial Neural Network (ANN) techniques to improve the accuracy and reliability of road selection during map-matching (MM) computation. MM positions provided by low-cost GPS receivers have great potential when integrated with hand-held or in-vehicle Geographical Information System (GIS) applications, especially those used for tracking and navigation, on path and road networks. The applied modular neural network (MNN) approach is using a suitable road shape indicator to incorporate different road shapes for local ANN training. MNN test results indicate good potential for the method to provide a significant improvement in MM and positional accuracy over traditional methods. Further results and conclusions of this on-going research will be published in due course.

Keywords: GIS, GPS, Artificial Neural Networks, Map Matching, Location Based Services.

1 Introduction

Accurate and reliable position determination is a vital component of information systems that are location dependent. These types of systems include in-car vehicle navigation systems, mobile phones or personal digital assistants performing location based services (LBS) and many others. Various methods are used to provide a position, many dependent on the use of the Global Positioning System (GPS) [1]. Currently, GPS, developed and maintained by the U.S. Department of Defense, is the only fully operational Global Navigation Satellite System (GNSS) available. Such systems utilise a number of satellites placed in earth orbits for terrestrial point positioning. GPS currently consists of 29 operational satellites, where at least four satellites have to be in view for simultaneous observation to obtain a three-dimensional receiver position using computed distances (ranges) between satellites and receiver. The fourth satellite is needed as there is a time difference to be considered between the satellites' clocks and the clock of the GPS receiver. GPS positioning errors occur from the cumulative effects of error due to the receiver, satellites and atmosphere, and may also be due to US military intentional accuracy limitation. With the autonomous European Navigation Satellite System "Galileo" expected in 2010, an opportunity of a joint system "GPS+Galileo" with more than 50

J.D. Carswell and T. Tezuka (Eds.): W2GIS 2006, LNCS 4295, pp. 76–89, 2006.

satellites will provide many advantages for civil users in terms of availability, reliability and accuracy. The number of GNSS satellites will further increase when the Russian GLONASS system is upgraded (ongoing).   According to the U.S. government's 2001 Federal Radionavigation Plan [2], low cost GPS provides an average positioning accuracy of 13 meters horizontally and 22 meters vertically, 95 percent of the time. A relatively recent report about the current capacity of civil GPS shows that the accuracy of a stand-alone GPS receiver such as a simple hand-held device might often be as good as 5 to 7 meters horizontally and 8 to 9 meters vertically [3]. In order to improve the accuracy of positions provided by GPS additional correction information may be used, such as Differential GPS (DGPS), or other sensors to enhance position reliability, such as a digital compass, gyroscope etc. DGPS is based on the concept that the errors in GPS position computation at one location are similar to those for all locations within a given (local) area.   Errors affecting the measurement of satellite range at a known location can be used as a correction to completely eliminated or at least significantly reduce error at an unknown location in the same local area.

A great many of applications of these location dependent systems are in use on roads and footpaths, e.g. intelligent transport systems (ITS) inside vehicles, personal navigation systems or LBS, often using hand-held GPS incorporated in a Personal Digital Assistant (PDA) or mobile phone. Generally travel on road or footpath, i.e. on a transport network (TN), may provide computer algorithms with digital information that can be used to correlate the computed system location with a digital map TN. This is known as map matching (MM) [4]. MM techniques vary from those using simple point data, integrated with optical gyro and velocity sensors [5] [6], to those using more complex mathematical techniques such as Kalman Filters [7] [8]. Existing literature revealed that a key function of any MM algorithm is to identify the correct road segment among the candidate road segments, since one incorrect match can lead to a sequence of incorrect matches. Particular attention has to be paid to topological aspects of the road network as well as matching processes at intersections (since most route changes occur there) and algorithm validation in complex route structure environments such as in built-up urban areas. Research into GPS/Odometer Integration using MM for a Local Transport Bus Company showed that GPS alone will often not meet requirements of reliability and positional accuracy, especially in urban areas, due to satellite masking by buildings and severe GPS signal multipath [9]. A specific MM algorithm using low cost stand-alone GPS has been developed to augment point position computation. This method, called Map Matched GPS (MMGPS), uses geometric information derived from large scale digital mapping, and digital terrain models (DTM) for height aiding [10] [11]. MMGPS tracks a vehicle along all potential road segments, which match its travel trajectory, defined by a time series of GPS derived coordinate positions.   In most cases, road segments are quickly filtered from the set of potential segments, until the correct road segment is identified. Other research to improve MMGPS includes an investigation of interpolation methods and accuracy of height data, which is described in Li et al. [12], and investigations using map intelligence and network analysis [13].

Research into the use of Artificial Neural Networks (ANNs) for road selection was successfully initiated by Winter [14]. The results of this research showed that MM can be effectively augmented with the aid of ANNs [15]. The initially created ANN indicated great potential for improving position accuracy for situations with a reduced number of satellites in view and poor satellite-receiver geometry, e.g. urban canyons, woodland areas or rugged terrain. ANNs are able to select the correct road using numerical inputs derived from the GPS receiver trajectory and the geometry of road centreline network.

Why use ANNs for augmentation of map-matched GPS positioning? A lot of research is on-going using Inertial Navigation Systems (INS), such as gyrocompass, odometer or flux gate compass, to improve GPS positioning reliability, since GPS satellites can easily be obstructed by high buildings, especially in cities, as well as by vegetation and terrain in rural areas. Disadvantages of INS are expense, reliance on other instruments or external correction and system complexity. Furthermore, with INS there is usually an accumulating drift error emerging over time. In comparison, ANNs once trained provide a high performance in the application phase for real-time applications [16]. Although, costs occur when high volumes of sample data have to be observed for ANN training. The inherent unreliability existing in these types of system, both ANN and INS based, typically occurs in situations where vehicles are moving from one road to another and the route is complex (multiple road junction), for example at a roundabouts, where the path scenario is changing. This problem is addressed with the use of a modular neural network (MNN) technique using a suitable road shape indicator to incorporate different ANNs for different road shapes for local expert (ANN) training.

2 Road Shape Indicators

This research deals with designing and developing a MNN technique that autonomously chooses the appropriate expert (ANN) from a number of locally trained ANNs, based on road shape, e.g. 90 degree bend, straight road. The intention was to create one local expert for various transport network (TN) road shapes, since initial work showed difficulties to optimize for different such road shapes in one ANN [15].

In order to correctly map-match the GPS positions, a decision about the correct road can be difficult, especially in any TN scenario, where:

> ➤ several roads meet or cross each other
> ➤ a car or moving person is going from one road to another
> ➤ two roads are in close proximity of each other and are parallel, or almost in parallel.

The different TN situations can change rather quickly, which makes it difficult to differentiate between the TN categories in order to choose the correct local ANN for road selection. This was a key challenge in this research. Which parameter will provide the best differentiation of varying TN road geometry? Considered mathematical measures that indicate TN segment curvature or profile were parameters derived from mathematical computations that are dependent on TN geometry, such as a fractal value of a line (digital TN segment). One such parameter is a dilution of precision (DOP), which is a measure derived from the least squares estimation (LSE) mathematical technique. The Correction Dilution of Precision (CDOP) is a DOP of a sequence of GPS position error vectors, derived during MMGPS map matching computation.

The Correction Dilution of Precision, now protected by a UK patent [17], has proven to be a very good parameter for indicating road geometry. CDOP is calculated using a sequence of GPS positions and road geometry. However, CDOP is a function of road geometry only [10], which can be computed using a mathematical formula with dependant variables of road centreline direction cosines and the number of coordinate positions used in the calculation.

$$\text{CDOP} = n^{-\frac{1}{2}}\left(\overline{\sin^2\phi\,\cos^2\phi} - \overline{\sin\phi\cos\phi}^2\right)^{-\frac{1}{2}}$$

Where n *is* the number of coordinate positions used and $\emptyset$ is the road segment bearing. As part of this research, CDOP was obtained using two techniques; Ordnance Survey road centreline coordinates using the theoretical formula above, and as a by product of the least squares GPS error vector estimation while map matching GPS point positions. Table 1, adapted from Blewitt et al. [10], displays CDOP for known values of $\emptyset$.

Table 1. Theoretical CDOP

Road Geometry	Correction Dilution of Precision, CDOP
Instant bend, angle α	$2/\sin\alpha\sqrt{n}$
Instant bend, 90°	$2/\sqrt{n}$
Instant bend, 45°	$2.8/\sqrt{n}$
Instant bend, 20°	$5.8/\sqrt{n}$
Instant bend, 10°	$11.5/\sqrt{n}$
Smoothest curve, α	$2/\sqrt{\left(1-\sin^2\alpha/\alpha^2\right)n}$
Smoothest curve, 90°	$2.6/\sqrt{n}$
Smoothest curve, 45°	$4.6/\sqrt{n}$
Smoothest curve, 20°	$10.0/\sqrt{n}$
Smoothest curve, 10°	$19.9/\sqrt{n}$

Based on the CDOP definition, the GPS error ceases to be a dominant error source when CDOP ≤ 1. This can be achieved by using only four GPS epochs when a vehicle drives along a right-angled bend. As more measurements are introduced, CDOP approaches 0, which implies that positioning is as good as using a perfect DGPS system.

Extensive tests were performed using CDOP as a new technique for categorising TN geometry within the research of improving map-matching. GPS observations were taken in a moving car on roads with different geometry, such as straight roads, smooth curves and instant turns. An example of the change in computed CDOP

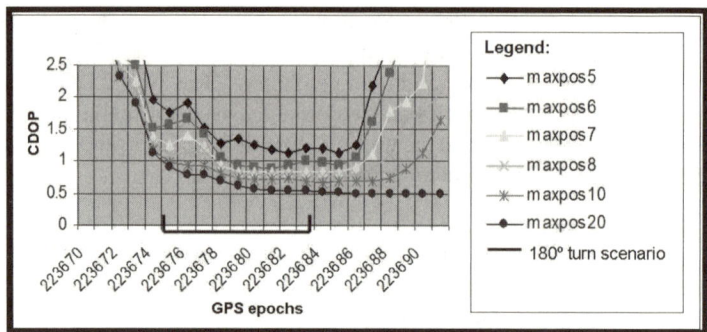

Fig. 1. CDOP during a typical 180° turn

valuon, whon a vohiclo trajectory desciibes a 180 degree turn, is shown in Fig. 1. The computation of CDOP can vary depending on the number of previous positions used in the LSE. Hence, CDOP values calculated with 5, 6, 7, 8, 10 and 20 previous positions (maxpos) were analysed. From Fig. 1, it can be noted that the more previous positions used in the computation and the higher the degree of turn (TN geometry), the lower the resulting CDOP value.

Comparing the computed values with the theoretical CDOP (see Table 1), the following conclusions can be drawn:

➢ The higher the degree of turn (road geometry), the more CDOP approaches its theoretical value.
➢ The more previous positions (maxpos) are used in the computation, the closer CDOP approaches its theoretical value.
➢ The fewer previous positions are used in the computation, the more quickly CDOP will indicate a change in road geometry.
➢ These conclusions lead to the approach of calculating CDOP using two different numbers of previous GPS positions, one in order to get an indicator as close as possible to the theoretical CDOP value, and another which detects a road geometry type as quickly as possible.

Table 2. Road Shape Ranges vs. CDOP Value Ranges

Road shapes ranges	Empirical found CDOP values ranges		Theoretical CDOP value ranges	
	Maxpos10	Maxpos5	Maxpos10	Maxpos5
180° - 70°	0.6 – 1	1 – 2	0.63 - 0.99	0.89 - 1.39
70° - 30°	1 – 2	2 – 4	0.99 - 2.13	1.39 - 3.01
30° - 10°	2 – 6	4 – 10	2.13 - 6.29	3.01 - 8.89
< 10°	> 6	> 10	> 6.29	> 8.89

Statistical analysis of numerous datasets identifies the following two CDOP conditions, which were considered in the proposed MNN approach when choosing a specific ANN for a particular road geometry type:

1. Using five previous vehicle positions (maxpos = 5) in the LSE is the best way to indicate the start and end of each road geometry type, Fig. 1.
2. Using ten previous vehicle positions (maxpos = 10) produces computed CDOP values, which more closely approach theoretical CDOP values, Fig. 1. This value of maxpos creates the least number of outliers in any road geometry type.

Table 2 shows that the empirically found CDOP value ranges reasonably match the theoretical CDOP values. Based on these CDOP value ranges together with visual inspection of many vehicle trajectories (GPS positions) as well as experiments on ANN training for different road geometry, four road shape categories were defined (see Table 2). These categories were used for the proposed modular neural network approach.

3 Proposed Modular Neural Network Technique

Artificial Neural Networks (ANNs) are information-processing devices composed of highly interconnected processing elements, which can learn by example to perform a generalisation. If it is known in advance that a set of training cases may be naturally divided into subsets that correspond to distinct subtasks, interference can be reduced by using a system composed of several different "expert" networks (ANNs) plus an indicator that decides which of the experts should be used for each training case. This is known as modular neural network (MNN) approach and is shown in Fig. 2. The input space can automatically be partitioned into regions so that each local expert takes responsibility for a different region. The proposed MNN technique is using CDOP as a road shape indicator to modularize ANN input into four road shape categories based on CDOP value ranges (see Table 2).

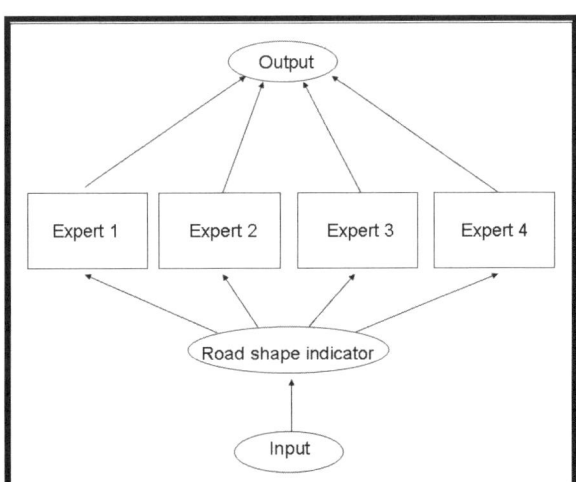

Fig. 2. Modular Neural Network Approach

The implementation of a MNN required the collection of a large number of GPS data for training and testing purposes. The data should cover all the different cases that the network may encounter in the application phase, i.e. road geometry types. The GPS data, used for ANN training and testing, consisted of about 9000 epochs (one second interval GPS positions) of all types of expected geometry in all possible road shape scenarios, such as sharp turns, smooth curves, straight roads, roundabouts, slip roads, parallel roads and junctions. These single frequency C/A code GPS positions have positional accuracies typically between 5m and 15m. Therefore, in complex vehicle route scenarios, such as roundabouts, compound road junctions and multiple slip roads, identifying the correct road on which a vehicle is traveling is not usually a trivial task.

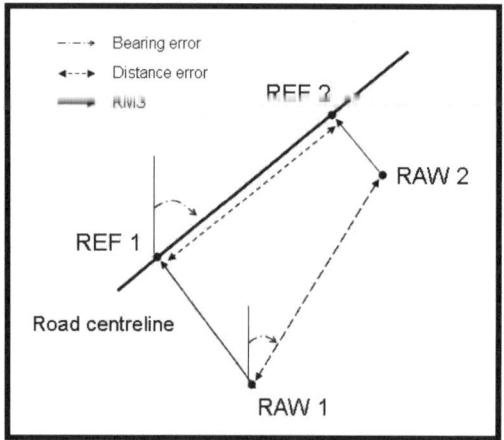

Fig. 3. Mathematical Relationships

The input parameters for ANN training were directly chosen from the computed relationship of the digital road geometry of road centrelines and the GPS receiver trajectory using MMGPS with an adjusted road reduction filter that did not eliminate any potential roads, except those that were clearly incorrect. This allowed the ANN to select the correct road, and not MMGPS. The mathematical relationships are demonstrated in Fig. 3, where for example, the bearing error is the difference between the RAW bearing, derived from the last (RAW1) to the current position (RAW2), and the bearing on the road centreline. These relationships (for bearing error, distance error, acceleration error and RMS) provide a direct correlation between the input parameters and the decision on the correct road (ANN output). Other input parameters, such as the GPS second, the road ID and the number of satellites, were identified empirically in experiments showing a positive effect on the ANN results. The local experts were trained as supervised networks, requiring input and corresponding output samples. With the input signal of the training data and the output decision, the ANN can be trained to create its own internal representation of relationship between input and output (decision rules). The decision on the correct road (expected ANN output) as well as the splitting of GPS data into different road shapes ranges was realised in an automated process. With the help of Visual Basic for Applications (VBA) different steps of automation were built into Excel spreadsheet macros.

4 Supervised Artificial Neural Network Training

An artificial neural network (ANN) has to be trained to process the input data in the required way. Supervised training is an iterative process and operates by adjusting the network weights in small steps so that network behaviour converges on the desired output (minimising the error between the desired and the computed unit values). Complex problems may require hundreds of thousands of training cycles. [18] ANN training and testing was performed using NeuralWorks Professional II/Plus by NeuralWare, providing a sophisticated environment with appropriate ANN simulations for this application.

Although not the focus of this research, a variety of different experiments on finding the best possible way of training the ANNs, with the given input data, were carried out before final ANN training and testing:

- Determining empirically how many hidden layer and nodes are necessary
- Testing the use of a radial basis function network (RBFN) in comparison to a back propagation network
- Experimenting with different input parameters
- Additional experiments on finding the appropriate number of local ANNs
- Training good data (without ANN output errors) and bad data (with ANN output errors) separately
- Experimenting with input parameters and CDOP processed in MMGPS with different numbers of previous positions (maxpos5, maxpos10) in the GPS trajectory.
- Determining how many input samples are optimal for local ANN training

Eventually, the back propagation network was chosen for local ANN training, as it is known as the most common supervised network [19], and RBFN did not prove to be a better network structure in this application. Back Propagation is a traditional supervised training algorithm using gradient descent for error minimisation [20]. For ANN activation, a sigmoid function was used, since it is the one that has been used most often successfully in Multi-Layer-Perceptrons [21], and a non-linear relationship between input and output exists. The Extended Delta-Bar-Delta learning rule [22] was the applied learning method, in which the problem of local minima, typical for a gradient descent algorithm such as back propagation, is tackled. A number of seven input parameters was finally chosen for ANN training (bearing error, distance error, acceleration error, RMS, GPS second, road ID and number of satellites). The network topology consists of one hidden layer with 15 hidden nodes, which complies with Kolmogorov' theorem [23]. The input space was normalised to the network range of [0, 1], which helps when using ANN parameters with different value ranges. The output network range was decided to be [0, 1], since the expected ANN output was prepared to be 0 for the incorrect road or 1 for the correct road. With the aid of NeuralWorks' SaveBest option overtraining was avoided, applying 2/3 of the training data for training and 1/3 for simultaneous testing. In a separate testing phase, the trained ANNs were tested on unseen test data. A summary of all ANN specifications for final ANN training is listed in Table 3. Four local experts were trained, that is, one ANN for each road shape range (see Table 2). The training data were automatically split into road shape ranges using CDOP as road shape indicator calculated with 10 previous positions (maxpos = 10) in the LSE of error vector (in MMGPS).

Table 3. Final ANN specifications

ANN specifications	
Network model	Back propagation network
Activation function	Sigmoid function
Learning rule	Extended Delta-Bar-Delta
Number of Inputs, n	7
Number of Hidden Layers	1
Number of Hidden Nodes, 2n+1	15
Number of Outputs	1
Input Network Range	[0,1]
Output Network Range	[0,1]
Initial Weights	Random
Termination Criteria (used in SaveBest)	RMS error of output layer
Performance Measure	ANN output errors in GPS epochs

The ANN training results are shown in Table 4. It can be seen that the number of training data with sharp turns and smooth curves (about 350 GPS epochs) is much smaller than the number of training data with almost straight and straight roads (between 600 and 900 GPS epochs), a typical distribution of geometry type along a road network. This distribution leads to a lower training time for ANNs trained with sharp turns and smooth curves and a better training performance with less ANN errors in the training data than the ANNs trained for almost straight and straight roads. Another reason for the latter is that map matching (MM) algorithms in general, are known to perform better for turns and curves, due to the difficulty of differentiating parameters for straight roads. Hence, the ANN input parameters produced in MMGPS are clearer for turns and curves.

Table 4. ANN Training Results

ANN Type	Epochs	ANN Errors	Errors All epochs
Sharp turns (180°-70°)	351	48	14%
Smooth curves (70°- 30°)	331	38	12%
Almost straight roads (30°- 10°)	606	114	19%
Straight roads (< 10°)	890	212	24%
All road shapes	2176	404	19%

Table 4 displays the results of the ANN training for the different local experts, defined by road shape (ANN Type). The total number of epochs (Epochs) is the number of times a GPS position was recorded. ANN errors are the number of epochs when a particular ANN did not select the correct road segment. The final column displays these errors as a percentage of epochs when the correct road was not selected. The last row of Table 4 displays the results for an all-in-one ANN (All road shapes), trained for all 2176 epochs, regardless of road shape.

The results of the training (Table 4) indicate immediately that basic ANN techniques are of potential benefit for road identification, when low cost GPS positioning is used to track a vehicle. However, the final ANN performance cannot be concluded from the results of ANN training. It requires an independent testing phase using unseen data.

5   Local ANN Performances

The trained ANNs were tested against new test data. This data is previously unseen by the ANNs, but the expected outcome (correct response) is known for each epoch. This testing phase is a way of determining how well the ANN learned the relationship between input and output and can perform generalisation [24]. In this testing the network weights are fixed and not further updated. To find out how often an ANN is giving the correct response, the ANN output values are compared with the expected output values. Ideally the network should produce the same output as the expected output, which is 1 for a correct road centreline and 0 for an incorrect road. But ANNs rarely give the exact result, which is desired [24]. In this application, the ANNs are presenting their outputs in the range of [0, 1]. Hence, the largest ANN output value in each epoch (GPS second) should point at the input sample for the correct road centreline (with the expected output = 1). The quality measure for the ANN performance is the number of ANN errors, which is the number of times (epochs) when the ANN was not able to select the correct road. This number of errors for the different single ANNs is shown in Table 5, together with the number of epochs and percentage errors when there is more than one road in the input data at a particular epoch. It should be noted that the percentage of epochs with more than one road (Epochs Roads>1), which is displayed in brackets for all ANN types, is a general indicator of the complexity of the road network with that road shape. The results presented in Table 5 display success rates for selecting the correct road ranging from 92% for well defined road geometry (sharp turns), to 86% for smooth curves. These small differences in performance of the local ANNs can most probably be attributed to natural variability in the relationship between the true car trajectory and the digitally recorded road centreline data used. These results in Table 5 would seem to indicate that a modular neural network (MNN) consisting of two local ANNs would provide the best results; one ANN for sharp turns and another for all other road geometry. For comparison purposes the all-in-one ANN for all road shapes was also tested and the results are shown in the last row in Table 5. Its performance is only slightly inferior to the local ANNs.

Table 5. ANN Test Results

ANN Type	All Epochs	Epochs Roads > 1	ANN Errors	Errors Roads > 1	Errors All epochs
Sharp turns (180° - 70°)	385	206 (54%)	29	14%	8%
Smooth curves (70° - 30°)	664	333 (50%)	94	28%	14%
Almost straight roads (30° - 10°)	2027	696 (34%)	192	28%	10%
Straight roads (< 10°)	3511	1281 (37%)	428	33%	12%
All road shapes	6587	2516 (38%)	772	31%	12%

Combining the results from the local ANNs provides an idea of how well the proposed MNN method is behaving in all road shapes. As can be seen in Table 6, the percentage of ANN errors for all road shapes is 30%. That is, for about two thirds of GPS epochs when a decision on the correct road was necessary the proposed MNN generated the correct output. The overall performance indicates an error of only 11% considering all epochs, which is a success rate of 89% for the correct road found.

By way of comparison the same 6587 epochs were processed using the original map matching software, MMGPS. MMGPS failed to select the correct road in 20% of all epochs, either selecting the wrong road, or not filtering out sufficient roads. Therefore, the MNN solution provides a 50% improvement over MMGPS. That is, MMGPS fails to identify the correct road twice as many times as the MNN approach. This ANN result will provide a significant augmentation for this map matching algorithm (MMGPS).

Table 6. MNN Results

Combined test results for all road shapes	
All Epochs	6587
Epochs Roads > 1	2516
ANN errors	743
Errors Roads > 1	30%
Errors All epochs	11%

6 Current and Future Work

To be able to evaluate the quality of the proposed system, a comparison to available existing systems with the same function, i.e. identifying the correct road on which a vehicle is travelling, is part of current work. The three following systems were chosen to assess the performance of the proposed MNN approach:

1. The test-bed application MMGPS, which is using a map-matching algorithm with the original RRF parameter thresholds to find the correct road.
2. A single ANN created for all road shapes to select the correct road based on initial investigations.
3. A MNN using a gating network available as simulation in NeuralWorks.

Initial results of the comparison to MMGPS (1) and the single ANN for all road shapes (2) were given in previous section. The final results of this comparison phase and the analysis of the results will be presented in due course. For comparison purposes, a similar complex ANN software called NeuroSolutions by NeuroDimention will also be tested in future research. Additionally, the intention is to experiment with more modern techniques such as Support Vector Machines [25] for the purpose of GPS positioning augmentation. The proposed MNN technique is aimed to be implemented in MMGPS to improve its map-matching algorithm. It could potentially be applied in situations, where MMGPS unable to filter out sufficient roads in order to identify the correct road.

7   Conclusions

Research has been focused on designing and developing a Modular Neural Network (MNN) technique for road selection, which autonomously chooses the appropriate Artificial Neural Network (ANN) from a number of locally trained ANNs based on suitable indicators. The indicator used in this research is the Correction Dilution of Precision (CDOP). This is a new technique for differentiating Transport Network (TN) geometry within the research of improving map-matching. Extensive tests on CDOP were performed to empirically find four appropriate road shape ranges based on CDOP values ranges. About 9000 GPS positions of a moving vehicle were observed and prepared for ANN training and testing. The GPS data were processed in a map matching application for GPS positioning (MMGPS) to provide the seven chosen ANN input parameters – GPS second, road ID, number of satellites, distance error, bearing error, acceleration error and RMS (residuals from LSE of GPS error vector) – as well as the road shape indicator CDOP for every potential road segment. The choice of input parameters was mathematically and empirically justified. An automated process was deployed to make a decision on the correct road for every GPS epoch. For the proposed MNN approach, the observed GPS positions were automatically split into road shape categories using CDOP.

Relatively basic supervised ANN training methods, such the method of back-propagation of error, were successfully applied for local ANN training, while being aware of the availability of more sophisticated and up-to-date learning methods. The locally trained ANNs for each road shape category were tested with success rates of approximately 92% to 89% based on the number of epochs with more than one road available. The ANN for sharp turns performed well and gave best results (92% success rate). The performance of the proposed MNN was derived from a combination of the test results of the locally trained ANNs. For about two third of GPS epochs when a decision on the correct road was necessary the proposed MNN generated the correct output, which is twice as good as the road reduction filter in MMGPS, over the exact same epochs.

References

1. Hoffmann-Wellenhof, B., H. Lichtenegger, and J. Collins, *Global Positioning System - Theory and Practice*. Fifth, revised edition ed. 2001, Wien/New York: Springer.
2. US Department of Defence and US Department of Transportation, *Federal Radionavigation Plan*. 2001.
3. Tiberius, C., *Standard Positioning Service - Handheld GPS Receiver Accuracy*, in *GPS World*. 2003. p. pp. 46-51.
4. White, C.E., D. Bernstein, and A.L. Kornhauser, *Some map matching algorithms for personal navigation assistants*. Transportation Research Part C: Emerging Technologies, 2000. **8**(1-6): p. 91-108.
5. Kim, J.S., J.H. Lee, T.H. Kang, W.Y. Lee, and Y.G. Kim. *Node Based Map Matching Algorithm For Car Navigation System*. in *29th ISATA Symposium*. 1996. Florence.
6. Mattos, P.G. *Intelligent Sensor Integration Algorithms for Vehicles*. in *Proceedings of ION GPS-93, Sixth International Technical Meeting of Satellite division of the institute of Navigation*. 1993. Salt Lake City, Utah.
7. Levy, L.J., *The Kalman Filter: Navigation's Integration Workhouse*, in *GPS World*. 1997. p. pp. 65-71.
8. Scott, C.A. *Improved GPS Positioning for Motor Vehicles Through Map Matching*. in *ION GPS-94, Seventh International Technical Meeting of the Satellite division of The institute of Navigation*. 1994. Salt Lake City, Utah.
9. Taylor, G., C. Brunsdon, J. Li, D. Steup, and M. Winter. *A Testbed Simulator for GPS and GIS Integrated Navigation and Positioning Research - Bus Positioning Using GPS Observations, Odometer Readings and Map Matching*. in *12th International Conference on Geoinformatics*. 2004. Gävle, Sweden.
10. Blewitt, G. and G. Taylor, *Mapping Dilution of Precision (MDOP) and Map Matched GPS*. International Journal of Geographical Information Science, 2002. **16**(1): p. pp. 55-67.
11. Taylor, G., G. Blewitt, D. Steup, S. Corbett, and A. Car, *Road Reduction Filtering for GPS-GIS Navigation*. Transactions in GIS, 2001. **5**(3): p. pp. 193-207.
12. Li, J., G. Taylor, and D. Kidner. *Accuracy and reliability of map matched GPS coordinates: dependence on terrain model resolution and interpolation algorithm*. in *GISRUK 2003*. 2003. London: City University.
13. Taylor, G., J. Uff, and A. Al-Hamadani. *GPS positioning using map-matching algorithms, drive restriction information and road network connectivity*. in *GISRUK*. 2001. Pontypridd, Wales, UK: University of Glamorgan.
14. Winter, M., *Application of Artificial Neural Networks to Map Matching for GPS Navigation*, in *Institute of Photogrammetry and Remote Sensing*. 2002, Dresden University of Technology / Germany and University of Glamorgan / UK. p. 102.
15. Winter, M. and G. Taylor. *Modular Neural Networks for Map-Matched GPS Positioning*. in *IEEE Web Information Systems Engineering Conference 2003 - Wireless Geographical Information Systems*. 2003. Rome.
16. Stonham, T.J. *Neural Networks - How they work -Their strengths and weaknesses*. in *Neural networks, neuro-fuzzy and other learning systems for engineering applications and research*. 1994. Institute of Civil Engineers, Saltford University: DRALL.
17. Taylor, G. and G. Blewitt, *Intelligent Positioning: GIS-GPS Unification*. Book. 2006: Wiley. ISBN: 0-470-85003-5.
18. UK Department of Trade and Industry, *Best practice guidelines for developing neural computing applications - Overview*. 1994, consortium of Touche Ross, AEA Technology and EDS-Scicon.

19. Hecht-Nielson, *Neurocomputing*. 1990, New York: Addison Wesely.
20. Picton, P., *Neural Networks*. 2000, New York: Palgrave. 0-333-80287-X.
21. Stergiou, C. and D. Siganos, *Neural Networks*. 1996.
22. NeuralWare, *Neural computing - A technology handbook for NeuralWorks Professional II/ Plus*. 2001a, Carnegie, USA.
23. Hecht-Neilson. *Kolmogorovo's Mapping Neural Networks Existence Theorem*. in *First IEEE International Conference on Neural Networks*. 1987. San Diego
24. NeuralWare, *NeuralWorks Professional II/Plus: Getting Started - A tutorial for Microsoft Windows Computers*. 2001b, Carnegie, USA.
25. Bennett, K.P. and C. Campbell, *Support Vector Machines: Hype or Hallelujah?* SIGKDD Explorations, 2000. **2**(2): p. pp. 1-13.
26. Blewitt, G. *Basics of the GPS technique: observation equations*. in *Geodetic Applications of GPS*. 1997. Nordic Geodetic Commission, Sweden.

Representational Issues in Interactive Wayfinding Systems: Navigating the Auckland University Campus

Alan Kwok Lun Cheung

School of Geography and Environmental Science, University of Auckland
10 Symonds Street, Auckland, New Zealand
ache051@ec.auckland.ac.nz

Abstract. This paper explores the effectiveness of different representational options in providing navigational cues to visitors to a site, in this case the City Campus of the University of Auckland. The research involved integrating the theory of wayfinding and spatial search behaviour with that of cartography and spatial visualisation, so as to identify appropriate approaches to designing and implementing an interactive system. Prototype navigation systems of different representations are produced for obtaining wayfinding performance data from participants. Ferom the results, advantages and disadvantages of the different systems are identified.

Keywords: Wayfinding, Exploratory cartography and interfaces, Mobile GIS.

1 Introduction

Wayfinding is one of the fundamental processes which human-beings perform in order to get to places, accomplish tasks or to gain knowledge in general. Wayfinding involves social interactions between wayfinders and informants. Traditionally, communication of route information was restricted, mainly via the medium of maps, textual descriptions and verbal descriptions.

The forms of communication have changed tremendously with the advent of new digital technologies. Communication of spatial information is no longer restricted to its mundane forms after the birth of portable Geographical Information Systems, which redefine the way geographical information can be communicated. Face to face verbal communication about experiences with certain spatial areas can now be augmented or replaced with intelligent interfaces that assist wayfinders to find the information they are after. In this study, we will examine different methods of representing wayfinding instructions by using current technologies, and evaluate their difference in performance within a case study for the University of Auckland City Campus environment.

The first objective of this research is the development of different wayfinding representational systems used for experimental deployment. The second objective is to evaluate the advantages and disadvantages of the systems produced from the previous phase, by conducting an examination of participants' travel through pre-planned routes within the campus area.

J.D. Carswell and T. Tezuka (Eds.): W2GIS 2006, LNCS 4295, pp. 90 – 101, 2006.

2 Definition of Wayfinding

Many people confuse wayfinding with route-finding. Some even consider them to be the same. However, wayfinding and route-finding are two different processes, both of which are required for the purpose of navigation. To clarify the difference, the purpose of wayfinding is to traverse the optimal route by using different cues, while route-finding involves devising optimal routes by evaluating different weighting factors such as distance, gradient etc.

A simple definition of wayfinding is the cognition of spatial information and decisions (not the movement) a person needs to perform to get from one place to the other. The term wayfinding was coined by Lynch [1] in 1960, it has been picked up and modified by researchers from various disciplines [2, 3]. Wayfinding was defined by Golledge [4] as:

'... the process of determining and following a path or route between an origin and a destination. It is a purposive, directed, and motivated activity. It may be observed as a trace of sensorimotor actions through an environment. The trace is called the route. The route results from implementing travel plan, which is an a priori activity that defines the sequence of segments and turn angles that comprise the path to be followed. The travel plan encapsulates the chosen strategy for path selection.'

Golledge had reinforced in his definition that wayfinding is not a task of finding an optimum route, but to travel the route by following the instructions (called travel plan in the definition). This definition is essentially the core part of wayfinding which all disciplines agree upon.

3 Knowledge in the World / Knowledge in the Head

Perception of reality inside a wayfinder's mind differs from person to person, and also from 'real' reality. Norman [5] classified perceptions and absolute reality into two divisions, where 'real' reality is 'Knowledge in the World' (KITW) and individual perception of reality is classified as 'Knowledge in the Head' (KITH). Most researchers explore the realm of what Norman calls as 'Knowledge in the Head' in the formation of cognitive maps or production of travel plans, in general deals with cognitive representation of spatial features in wayfinders' minds.

Norman did not examine the respective linkages between KITW and KITH. But KITW and the interface are in fact an integral part of the wayfinding process. It is especially important to look at this interface because it involves the cognitive process of transforming KITW into KITH. If this interface is well established (like in the travel guidebooks that include landmark locations, descriptions, etc.), where KITW can be transformed easily into understandable language of KITH, then the wayfinding process is most likely to succeed.

Krafft mentioned in his research that:

'Environmental knowledge is that which is acquired about the real world environment through any number of processes. Direct interaction would be the most obvious, but knowledge of the environment can be attained in many other ways. Media, in the form of written descriptions, pictures, drawings or other schematic representations, as well as verbal descriptions passed on by other individuals can also provide an adequate representation of the environment for purposes of spatial knowledge acquisition...' [6]

This definition demonstrates that the acquisition of spatial information is not monotonous, but could be done using different forms of media, or 'interfaces'. Geospatial information provided by the interface is an abstracted and recompiled version of reality. Not only it could provide 'real' information supplied by the spatial feature described, but it could also include metadata which is not embedded into the feature, for example history of a building or what an architect wants to express via architectural design. With this characteristic, a wayfinder who extracts information from KITW using an effective interface has an advantage over a wayfinder who only interacts directly with KITW, because more potentially useful data is available for wayfinding.

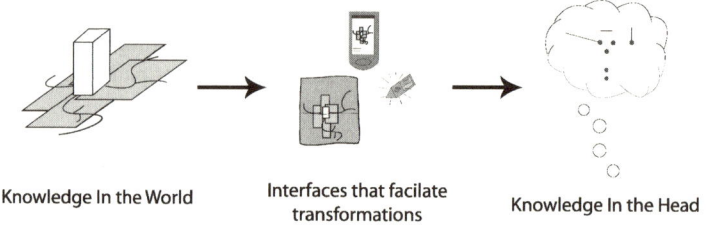

Fig. 1. How Knowledge in the World is related to Knowledge in the Head

4   Examining Wayfinding Process

As commented earlier, wayfinding performance must be differentiated from route traversing performance. Although route traversing time is a strong indicator of how effective a wayfinder travels, pure measurement of time however do not have the capability to accommodate different parameters that would affect performance of wayfinding. Measuring wayfinding performances, it is necessary to examine wayfinders' decision making process for solving wayfinding problems [7]. It is accepted that wayfinders make most of the decisions while passing landmarks, standing on intersections, etc.

For this purpose, a group of 30 participants were recruited on a voluntary basis, 15 female participants and 15 male participants, aged ranges from 13 to 30. Some of the participants were non-native English speakers. Most of the participants held an academic degree and five of them majored in subjects related to map interpretation and reading (Geography, Engineering, etc.). More than half of the participants had experience in navigating the university campus area.

The task given to the participants (hereby called "wayfinders") was to traverse four designated routes within the University of Auckland campus by following the instructions given by different route representation systems running on a PDA. Each of the sessions was recorded in video for the durations of the experiment. The video footage allowed capturing of wayfinders' verbal comments, gestures and any other actions performed by wayfinders. By isolating the traverse process from wayfinding process, it was also possible to verify the influence of the wayfinders' physique on their overall performance. It must be noted that effects of physical differences on results still exist, such as eyesight or body height that might be advantageous to some wayfinders over the others.

5   Selection of Study Area

The City Campus of the University of Auckland was chosen since it satisfies all of the requirements needed to achieve a homogeneous environment with the right level of spatial configuration complexity. The study area is located in the centre of Auckland City, and there is no discrete boundary because the campus is assimilated with the surrounding urban area. This study will focus on wayfinding inside the nucleus area because all spatial features are within a reasonable walking distance and the spatial configuration of the area is sufficiently rich in spatial features as references for the purpose of this study. Also given the adequate amount of prominent spatial features within the campus area, wayfinders would be able to navigate by following the instructions with minimal problem.

6   Selection of Representations

There is a wide range of representations that one could choose from to provide wayfinding instructions. However, the idea was to implement systems that are not obscured or too different from what normal people could understand and learn in a short time. The candidates chosen were interactive map, virtual environment, annotated images, textual instructions, verbal instructions, recorded video and navigation arrows powered by GPS. And from this list, three final representation methods were chosen based on the familiarity by people, difficulty of implementation and their relevance for providing wayfinding instructions on the fly during the wayfinding process. The three dimensional virtual environment was declined at an early stage because of the difficulties involved in developing applications that would run on a PDA. Textual instructions, although highly informative, requires some time to interpret, and would not be suitable to provide instructions on the fly. Recorded video, due to the limited quality of video obtainable using current equipment, was discarded as well. As a result, interactive map, annotated images and verbal instructions were chosen for the familiarity and flexibility they can offer.

The implementation of the representation systems in PDA environment required careful considerations. The verbal system is simple, because it only requires recording and playing of audio clips. For the other two methods, it is not as simple. Programming languages such as HTML, Flash + ActionScript, SVG + JavaScript and Processing (a derivation of Java) were considered. Despite the rough programming environment, lack of support and possible compatibility problems, SVG has been chosen over Flash due to the minimum budget of this study, and familiarity with ECMAScript used in SVG over ActionScript used in Flash.

7   Results and Discussions

The routes used in the experiment were designed so participants would face different wayfinding obstacles. Route 1 and 3 are standard length routes which led participants into underpasses and into buildings, thus utilization of the third dimension is ensured. Route 2 and 4 are exteneded length routes which incorporates long straight route segments, and in some cases open spaces. Difference in wayfinding performance and the related obstacles are discusses in the following sections.

Table 1. Average time participants spent at each waypoint performing wayfinding decisions

	Route 1	Route 2	Route 3	Route 4
Map-based Group	1.51s	1.57s	1.39	1.25s
Image-based Group	1.13s	1.78s	1.36s	1.66s
Verbal-based Group	1.38s	1.18	1.14s	1.66s

7.1   Map-Based System

Maps, in their various forms, are more established and accepted as a representational method compared to the other two methods used in the experiment. However, a legacy of traditional maps is the difficulty to represent real three-dimensional relationships in a map. With this respect, the main difficulty for wayfinders using the map system is the interpretation of three dimensional routes represented on a two dimensional map. As a result wayfinders often made wrong turns or complained about misrepresentation in the system. Features such as staircases and underpasses caused confusions before the wayfinders got used to the respective representations. Interestingly enough, wayfinders who made more mistakes were people who considered themselves to be familiar with the local environment. As suggested by Golledge [8], misinterpretation could be caused by the incorporation of own knowledge of the area with the instructions given by the map system, together coupled with the compromised representations; they were confused and made wrong decisions.

Wayfinders eventually became familiar with the twisted representations of stairs. Despite the fact they managed to interpret the relationship of "twisted route == vertical motion", their interpretation of such representation was not the same as the map maker perceived. Rather than treating the twisted route segments as a traversable path, wayfinders treated the segment as an embedded symbol of vertical movement, very much like other cartographic symbols which depict elevators or bus stops.

Apart from the problem regarding the third dimension, there were also a few other interesting issues. Firstly, wayfinders had to spend time reading the map before they started travelling. During this 'initialization process', wayfinders had to align themselves (or the map) so they could keep their position and orientation in-line with the real environment. This process was necessary even for participants who were very familiar with the university campus area, because they still needed to relate the geometric properties represented on the map which they were not familiar with, to the actual features in the familiar environment. However the initialization process only applied to the map users, and not to the users of other two systems.

Secondly, some wayfinders tend to forget very easily where their position is on the map. This finding accords with earlier research on wayfinding [6]. Relations between the map and the real environment generated by the participants while interpreting the map, are not solid, and can easily be interrupted as soon as the wayfinders stop communicating with the representation system. Whenever the wayfinder wants to initialize the wayfinding process, he/she has to re-localize both in reality and in virtual space. This re-localization is needed no matter how familiar the wayfinder is with the environment. In many instances locating oneself on a map is a trial and error process. Even though a wayfinder might have a detailed knowledge of the surrounding area, most of the features he/she can see would be diluted because maps provides generalised information to achieve clarity of representation by removing redundant data. Because the map only offers an abstracted and iconised representation of the real world, it is highly possible that the wayfinders could not find the necessary features to orient themselves.

Thirdly, most maps, apart from dynamically adjusting displays on location aware PDAs, are not oriented to the direction in which the wayfinder is facing. Any printed maps would include a north arrow. But without sound knowledge of the local space or a compass, the wayfinder might have little idea about their cardinal directions.

Wayfinders using other systems could suffer from the same problem. However, image and verbal systems rely on symbolic descriptions rather than geometric ones used in maps, therefore participants do not need to interpret as much information as map users do. Consequently, the initialization processes using image or verbal system was not evident in this experiment.

7.2 Image-Based System

The capability of the image-based system was affected by the visual depth of images, which in turn was affected by various practical constraints, such as definition of the camera lens, quality of the image and size of the PDA screen. Long straight route segments were particularly vulnerable to visual depth because the next waypoints were hardly visible and thus wayfinders could not see clearly where the route segment leads to and where to stop. The problem of visual depth had prompted the participants to develop strategies to counter it. One was to zoom into the image and get a more detailed view. And at later stages of the experiment, wayfinders discovered that blindly following the arrow as instructed without seeing the destination clearly usually worked fine. As a result, wayfinders spent less time at judging the next waypoint without making more mistakes than before.

A problem related to the 'visible dimension' of an image is the field of view. The problem aroused when wayfinders could not see the next waypoint on the current image because it locates was located outside the image. This problem did not cause much trouble for the participants in this study, but it shall be mentioned here to assist future work: Wayfinders were puzzled by arrows that only led them outside the current image, without a proper waypoint. Wayfinders were expected to follow the direction of the arrows and then forwarding into the next image. Instead, wayfinders preferred to know the location of the waypoint inside the current image, even if it was obstructed by foreground objects. As for the case for depth of view of images, it took some time before they were convinced that it was safe to carry out the instructions without seeing where the waypoints were located.

Apart from the quality of the images, continuity between images played a big role in creating a flow from waypoint to waypoint. In places where linkages between images were not well defined (e.g. the second image shows a scene completely different from the fist image), wayfinders had doubt whether they arrived at the correct location as depicted in the image. Problems also occurred when similar scenes taken at different locations were organized into consecutive sequence. This problem tended to happen in open space with lack of landmarks or when similar but different landmarks appeared on consecutive images. Since the general environment within the campus area is quite homogeneous, it was not surprising for wayfinders to misinterpret a building which bears a similar outlook as the one appeared on an image. An example from the experiment involved a set of identical looking bike racks. Two consecutive images were both showing a bike rack with a parked bike on the left hand side. Wayfinders who were not familiar with the area thought that the second image is a straight line continuation of the first, but in fact the view of the second image involved a third bike rack which is round the corner.

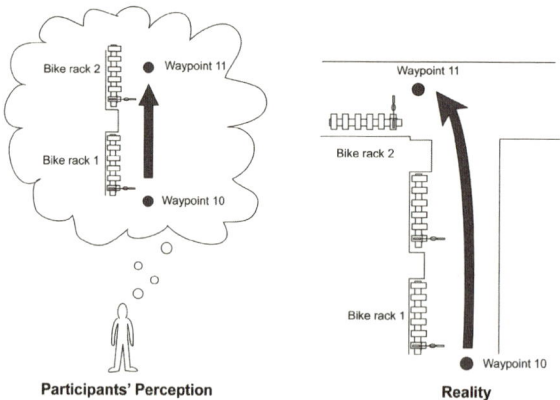

Fig. 2. The difference between perception and reality

Perhaps the most interesting finding from the image-based system experiment is that the shapes of the arrows used in the annotated images affected the participants' judgement of distances and turns that they needed to perform. In previous studies of navigation by arrows, the arrows were either made as non variating in width and perspective [9, 10]. In our experiment all arrows were implemented to lead the user from their vantage point up to the next waypoint (or to the edge of an image), therefore an indication of both distance and angle was conveyed. In several instances, the perspective of the arrows were not correct in relation to the visual depth of the image, because such requirement was not predicted when the system was produced. From comments made by wayfinders, we learnt that many of them judged travelling distances by looking at how the arrows behaved (e.g. arrow became thinner as they were "further away" from the user) (Fig. 3a). In other words, participants had treated the form of the arrows as they are embedded with geometric properties. As mentioned

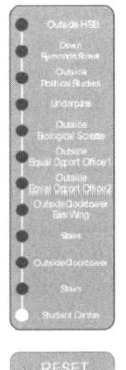

Fig. 3a. The arrow does not show the distance properly because its visual depth is different from the image **3b.** This image was taken from a perspective different from the participants' who supposed to come in through the doors.

before, wayfinders had since given up judging distances from the arrows and travel with only directional information obtained from them, and results had shown that their wayfinding performance was not affected.

Apart from the problem caused by the misaligned arrows, participants also commented that many of the images were taken from a perspective different from theirs. In order to match the perspective of view on the image, they needed to either move to where the image was taken, or to establish a relationship between their physical location and the location on the PDA display.

7.3 Verbal-Based System

The most controversial and different representation system in this experiment was the verbal based system. The most distinctive difference that affected wayfinding performance was the lack of visual representation. Without visual representation, wayfinders were not able to perform visual matching between the representation and reality, and it was expected that it would weaken the wayfinding performance. However, this prediction was proved to be incorrect, at least in this study. Our results had shown that the overall performance of the verbal system was significantly better than the other two.

Despite the fact that verbal cues provided less spatial information for users, this actually became an advantage when used as a wayfinding aid instead of pure spatial information provider because of the lack of distractions. Also since there was no visual feedback from the system, therefore no visual matching with the environment is needed, or more precisely, visual matching was not required. The conceptualization of wayfinding tasks and spatial features was the same for all systems tested. But since the amount of spatial objects and their relationships in verbal system was restricted to what was directly related to the route, therefore the conceptualization process was greatly simplified compared to visual systems. Although they had nothing to look at, they did however 'recite' the instructions while travelling the route. By repeating the instructions, not only they saved the time from replaying and listening to the instructions, but they also established linkages between the spatial features described in the instructions with the ones in the real environment. From the examination of video clips and written records, it has been found that wayfinders who recited the instructions seldom needed to listen to them again, while wayfinders who kept quiet had to replay the instructions every so often. Some instructions were replayed the most:

1) Instructions that included more than three manoeuvres
2) Instructions that included spatial features with a functional or ornamental name (e.g. Recreation Centre, Equal Opportunities Office, etc.)
3) Instructions that referred to a feature without signifying its direction in relation to the wayfinders' location

Contrary to the map system verbal instructions performed well in interpreting three dimensional movements. The Dimensionless property of verbal instructions means the necessity to have a vantage point is not as important. Another factor that minimized the effect of dimensional problems was the use of symbolic-oriented instructions, as opposed to geometric oriented instructions. Geometric instructions had an advantage of being more accurate and detailed, but were cumbersome at times. Symbolic instructions were vague. Instead of describing each angle and distance the users had to travel, they were instructed to follow landmarks. Thus performing wayfinding task using symbolic verbal system was partly an exploration for the users because they had to find and identify landmarks. Symbolic instruction worked well in this study mainly because, as many wayfinders commented, the plentiful number of 'tips' were given about landmarks in the campus area. In addition, most buildings were labeled anyway, and there was also a large number of sign posts that pointed at the direction of landmarks used in the route.

However, the verbal system was not without problems. For example, at one waypoint, wayfinders were given a seemingly simple instruction: 'Walk straight ahead and turn left at the first T-junction.' Apparently the word 'T-junction' was not understood by many of the wayfinders. Surprisingly, none of the wayfinders wandered off track. They later commented that they managed to decipher the meaning of the word because from their position until the end of the walkway, there was only one type of feature which is 'countable', traversable and was directly related to the route. Nevertheless, wayfinders might not be able to decipher unfamiliar terms if the route sections are located in an area with more complex spatial configurations. The example shows that verbal instructions are not as universal as maps and images, and to interpret verbal instructions, wayfinders must first understand the specific language that the instructions use. If the language ability of the wayfinder is limited, there is a great chance that one could not understand the instructions. However, it does not signify that the wayfinder lacks interpretation or wayfinding skills.

As mentioned before, verbal instructions omitted information, and efficiency of communication plummets if the spatial configurations lack landmarks for reference. Due to the limitation of human's short term memory and the requirement to fit instructions into a reasonable time frame, the amount of information that could fit into a single instruction was very limited. Not only features that were not related to the route segment were removed from the instructions but, in many cases features which direct connection to the route were also removed because instructions would otherwise become very clumsy. The results from the omission were mixed, and in many cases the effect of omission depended on how a person interpreted an instruction with omissions. This problem was evident when the instruction "Go up the stairs at on the left, and then walk in a straight line until Equal Opportunities Office." was given. The confusion arose because wayfinders did not know where the Equal Opportunities Office is, not even people who reported to be very familiar with the area. The reasons being that the office is not a distinctive building and the usage of the current office had changed several times. For the image-based system, when wayfinders were confused from the instructions, they could still find the 'place' because the spatial configuration depicted in the image was not restricted to prominent landmarks. Other features, such as sign post and trees also contributed to the formation of the local spatial configuration known as a 'place'. However, verbal system users were at a disadvantage because they did not have an interface which presents the local spatial configuration. In other words, if they could not find the office building itself, there was no way they could use other features to form the spatial relationships needed to find out which was the office building.

Another information omission problem was with the instruction: "Walk towards to the stone wall then go through the stone wall". Some wayfinders were confused about how the stone wall looks like and where the passage of the stone wall is located because they entered the waypoint at an angle different from originally anticipated. As a result they could not see the passage. Furthermore, without visual aids it was hard for the wayfinders to align to the ideal angle. Finally, the area around the stone wall could be considered as open space. The effectiveness of verbal instructions describing open space environments declines if a specific destination or task is not well defined due to the virtually infinite number of ways the wayfinders could enter and navigate in an open space environment.

It is interesting to note that information omission was also applied to many of other waypoints without misleading the wayfinders. In fact, wayfinders did not notice that the information had been omitted at all. This demonstrates that effective and strategic omissions do not affect wayfinding performance. However there are also situations where omission must be restricted in order to preserve at least some of the spatial configurations of a place.

In summary, verbal instructions omit more information than the other two systems used in this study. There is a possibility that omission of information actually decreased wayfinding performance, however since only one set of instructions were prepared for each route, it was not possible to test whether this postulation is true.

8 Conclusions

Results from the experiment have shown that many previous beliefs on wayfinding performances could potentially be different from originally anticipated. Different representational methods do perform differently under different environmental conditions. Although the verbal-based system was the most consistent system in this study, it does not mean it performs equally well in other environmental conditions. For example, map-based and image-based systems performed better in navigating the participants through open space or places that are lean in landmarks.

The following list presents characteristics of the systems examined and the findings from the experiments:

Table 2. Comparison table of the implemented systems

	Advantages	**Disadvantages**
Map-based System	• Rich in information	• Requires initialization • Requires longer cognition time • Problem in displaying three dimensional information in its current form
Image-based System	• Realistic description of spatial features • Spatial configuration is embedded • Geometric instructions can be embed into instruction arrows	• Vulnerable to landmark changes • Lack of visual depth as one of the main problems • Obstructions of foreground objects • Difficulty in maintaining a flow between images
Verbal-based System	• Abstract, Direct Instruction • No dimensional barriers • Can be geometry driven or task driven	• Efficiency plummets in open space or lean environment

In conclusion, the experiments presented in this research suggest that the effectiveness of a representation system depends on the nature of the wayfinding task. One should always remember that the best wayfinding representation system does not constitute as the best system for providing survey knowledge, because the nature of the two tasks is different. A map-based system certainly be preferred over other types of systems if the experiment was designed to examine the amount of information participants could extract from representation system, because it provides much more information than verbal instruction.

Acknowledgments

Thank You to my supervisor Prof. Pip Forer who provide helpful hints through out the project. I would also like to thank Mr. Igor Drecki, Mr. Graeme Glen, Miss Rebecca Lau, Dr. Patrick Laube, Mr. Chris McDowall, Mr. Tim Nolan, and Miss Paulina Wong for their inputs and encouragements.

References

1. Lynch, K.: The image of the city. Cambridge [Mass.] Technology Press, Cambridge (1960)
2. Passini, R.: Wayfinding in Architecture. Van Nostrand Reinhold, New York (1984)
3. Garling, T., Book, A., Ergezen, N.: Adult's Memory Representations of the Spatial Properties of Their Everyday Physical Environment. In: Coher, T. (ed.): Development of Spatial Cognition, Hillsdale Press, Erlbaum, New Jersey (1984)

4. Golledge, R.G.: Wayfinding behavior : cognitive mapping and other spatial processes. Johns Hopkins University Press, Baltimore (1999)
5. Norman, D.: The Design of Everyday Things. Doubleday, New York (1988)
6. Krafft, M.F.: A Neural Optimal Controller Architecture for Wayfinding Behavior. Computer Science and Psychology. Swarthmore College, Swarthmore (2001)
7. Stern, E., Portugali, J.: Environmental Cognition and Decision Making in Urban Navigation. In: Golledge, R.G. (ed.): Wayfinding behavior : cognitive mapping and other spatial processes. Johns Hopkins University Press, Baltimore (1999) 99-119
8. Golledge, R.G.D., Valerie; Bell, Scott: Acquiring Spatial Knowledge: Survey versus Route-Based Knowledge in Unfamiliar Environments. Annals of the Association of American Geographers **85** (1995) 134-158
9. Kray, C., Laakso, K., Elting, C., Coors, V.: Presenting Route Instructions on Mobile Devices. IUI'03, Miami, Florida, USA (2003)
10. Lu, Y. H., Delp, E.: An Overview of Problems in Image-Based Location Awareness and Navigation, Unpublished (2005)

A Hybrid Spatial Model for Representing Indoor Environments

Bernhard Lorenz, Hans Jürgen Ohlbach, and Edgar-Philipp Stoffel

University of Munich, Oettingenstr. 67, 80538 Munich, Germany
{lorenz, ohlbach, stoffel}@pms.ifi.lmu.de
http://www.pms.ifi.lmu.de/

Abstract. In this article we propose a hybrid spatial model for indoor environments. The model consists of hierarchically structured graphs with typed edges and nodes. The model is hybrid in the sense that nodes and edges can be labelled with qualitative as well as quantitative information. The graphs support wayfinding and, in addition, provide helpful information for generating human-oriented descriptions of an indoor (and outdoor) path.

Keywords: indoor wayfinding, indoor navigation, hybrid spatial representation, qualitative spatial representation, topology, graph, hierarchical graph.

1   Introduction and Motivation

Car navigation systems are becoming a more or less standard commodity nowadays. Since they are a mass product, the problem of navigating cars through large road networks has been well investigated and the solutions are mature. Much less well investigated is the problem of navigating pedestrians through airports, train stations, libraries, hospitals, supermarkets, etc. Unfortunately, it turns out that navigation problems in large buildings are quite different from the ones encountered in large road networks. The main reason for this is that the topological structure of buildings is much more diverse than the topological structure of road networks: Whereas roads are primarily one-dimensional structures with landmarks aligned along them, areas in buildings are really two-dimensional (as in floor plans), or, when multiple storeys are taken into account, even 2.5-dimensional structures [4].

The most efficient wayfinding [8] solutions use shortest-path algorithms in graphs. Therefore, we propose a graph representation also for indoor environments, yet the graphs are much more structured and enriched with extra qualitative and quantitative information assorted. Beyond, when considering two-dimensional structures not only adjacency has to be modelled, but also containment. We introduce a hierarchical graph structure in section 3.

The main characteristics of the graph model are:

J.D. Carswell and T. Tezuka (Eds.): W2GIS 2006, LNCS 4295, pp. 102–112, 2006.
© Springer-Verlag Berlin Heidelberg 2006

- the two-dimensional areas in buildings are partitioned into cells, and these cells are represented as nodes in the graph (Section 2). Doors and other passways which represent possibilities for persons to move from one cell to another are represented as edges;
- in order to facilitate hierarchical planning, there are different levels of abstraction in the graph (Section 3). For example, a storey in a building may be represented as a graph at a certain level, this entire graph being just a node in a graph at a higher level which stands for the whole building. The edges in the abstract graph connect the different storeys;
- the nodes and edges of the graph are labelled with hybrid information to support wayfinding as well as the generation of a human-understandable description of a path. Hence, we distinguish different types of nodes (Section 4). For example, rooms and corridors are both represented as nodes, but with different labels. As we shall see, it is quite useful to maintain a list of doors and windows in a room, all sorted by their angle against a fixed point of reference (Section 4.1). Corridors, on the other hand, are essentially one-dimensional structures for which it is useful to maintain the sequence of doors at the left hand side and the sequence of doors at the right hand side (Section 4.2).

The indoor model is described in more detail in the subsequent sections. However, we want to emphasise that the model is deliberately kept flexible. The node and the edge types as well as their labelling can be extended when it turns out that this is suitable for future applications.

2   Cell Decomposition

For buildings with simple rooms and corridors, that is to say rather small rooms (unlike, for instance, an entrance hall where hundreds of people fit in) and narrow enough corridors (not stretching over several parts of a building), there is a direct one-to-one mapping to a graph structure. Rooms and corridors are represented as nodes, and the passways between them as edges. In Fig. 1, where an extract from a blueprint of a university building is shown, such a graph structure is laid over the floor plan.[1] Two rooms which are connected by two or more doors have two or more edges between the corresponding nodes (like the entrance hall and the main corridor in Fig. 1).

However, strictly pursuing this naïve approach becomes difficult for larger buildings with large areas of open space, as for instance an airport. Following Bittner [1] we divide the free space C_{free} in this case into non-overlapping, disjoint cells C_r such that $C_{free} = \bigcup_r C_r \wedge \forall i \neq j : C_i \cap C_j = \emptyset$. Adjacent cells are connected by a link. The main corridor in Fig. 1 is actually split into several cells due to its length. Otherwise, impractical route descriptions like *"turn left to the main corridor and take the 32nd door on the right"* may result.

The sheer size of a room may be a reason to decompose it into cells. Other reasons have to do with concavity of rooms, or with the functionality of certain

---

[1] The stairs to the other two storeys were omitted for keeping the example simple.

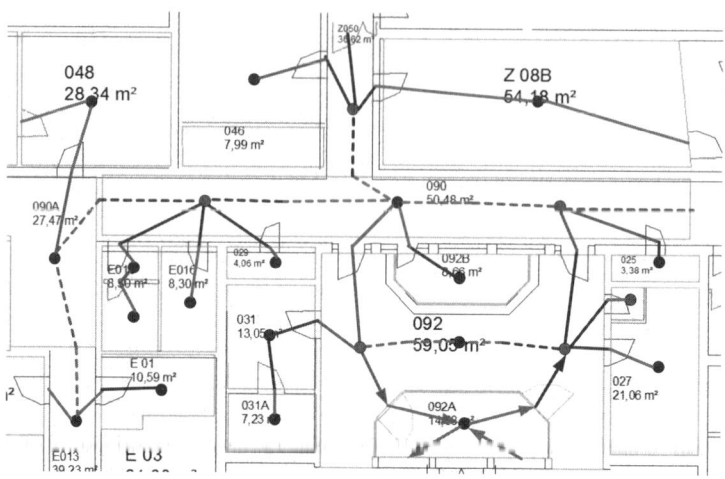

Fig. 1. Floor Plan Overlaid with Cell Centres and a Path System

areas in a larger open space. For example, an airport lounge may feature waiting areas, meeting points, areas in front of the different counters and security checks, passport control, etc. All of them serve a different purpose, and this must be represented in the graph.

Unfortunately, there is no obvious way to fully automate the cell decomposition. It has to be designed very carefully, taking into account the purpose of the different cells.

3 Hierarchical Graphs

If you are at the first floor of a large building, and you ask someone how to get to a particular room, the explanation may well start with *"Go to the third floor ..."*. What is behind this is a two-level (or, in general, multi-level) hierarchical model of the building. An example is depicted in Fig. 2. The upper hierarchy level consists of the storeys, and the lower level models the topology of each storey. In addition, the hierarchy shown in Fig. 2 also has an intermediate level which consists of wings. Navigation between different storeys usually consists of the steps *"go to the lift (staircase etc.)"*, *"go to the target storey"*, *"navigate the target storey"*. This is a typical case of hierarchical planning as it has been investigated in Artificial Intelligence for decades.

Our graph model supports hierarchical planning by providing hierarchical graphs. Each graph has a level (in the hierarchy) and an identifier. Graphs at higher levels can have as node labels the identifiers of graphs at lower levels. But this is not enough. There must also be a possibility to access graphs at higher levels from nodes of graphs at lower levels. This is done by classifying certain nodes of graphs at level n as "interface nodes" to graphs at level $n + 1$.

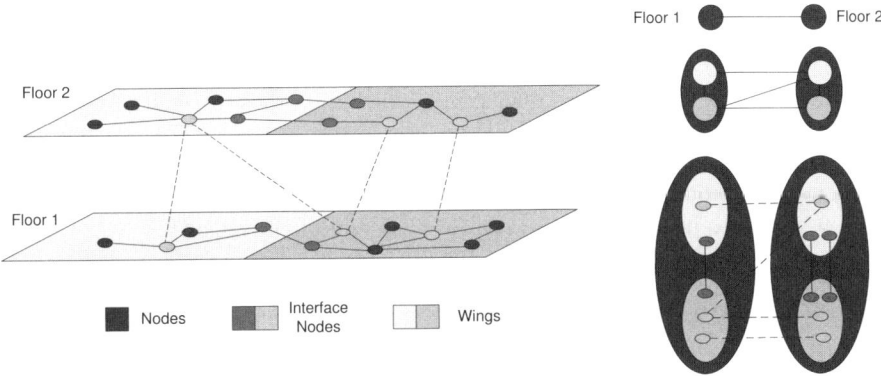

Fig. 2. Hierarchical Graph

Physically, these interface nodes may represent access points to staircases, lift doors, etc. (see Fig. 2).

The primary use for the graph hierarchy is of course the representation of different storeys in a building. Other use cases may necessitate the representation of different wings in a building (as in Fig. 2). Wings and storeys yield a hierarchy of three levels. If it makes sense to subdivide wings further, one may have four or more levels (see Fig. 3). On the other hand, there may also be further levels above the level of storeys. If we want to represent not only a single building, but, say, the whole campus of a university with many buildings, each building would be a node in a graph one level above the level of storeys.

A further use of hierarchical graphs can be the representation of areas which are contained within each other. As an example, consider the vegetable area in a hypermarket. The vegetable area may be subdivided into the area with the salad, the cucumbers, the carrots, etc. In the hierarchical graph model, we would have a node for the vegetable area at some level n, and this node refers to the graph of the salad, cucumber etc. areas at level $n-1$.

Model Element〈br〉Granularity	Graph	Vertex	Edge
Level 4〈br〉(coarsest)	City	Building	StreetNetwork
Level 3	Building	Storey,〈br〉*Staircase,*〈br〉*Elevator*	
Level 2	Storey	Wing,〈br〉Room, Corridor	Door,〈br〉Window,〈br〉*Ladder,*〈br〉Ramp,〈br〉Stairs
Level 1	Wing	Room, Corridor	
Level 0〈br〉(finest)	Room, Corridor	PartOfRoom,〈br〉PartOfCorridor	

Fig. 3. Relations between Hierarchy Levels and Graph Elements

The edges in the graph at the 'building level' represent walkways or streets. In the simplest case, such an edge contains solely the information that it is *possible* to get to another building. If we want more detail on *how* to get to this building, we must link the edge with another graph which describes the walkways and the road network. Therefore not only nodes of a graph at level $n + 1$ can represent graphs at level n, but also edges at level $n + 1$ can represent graphs at level n. The only difference is that an edge at level $n + 1$ must correspond to a graph at level n with two interface nodes, one for each end of the edge.

4   Node and Edge Types

Wayfinding by means of shortest path algorithms requires no more than a graph and a cost function. A simple cost function measures the geometrical distance between two places. More sophisticated cost functions can, for example, distinguish between lifts and staircases by making the staircases more "expensive". A minimum of semantic information is sufficient for this purpose. It turns out that the problem of wayfinding is considerably easier than the problem of describing an indoor path in a human-understandable manner. Humans use a combination of mostly qualitative information ("use the door *at the end* of the corridor") with little quantitative information ("take the *second* door to your left") for describing routes. Landmarks, which are very important in outdoor scenarios ("after passing by the church"), however, seem to be less important for indoor scenarios.

In order to support the generation of descriptions of a path through a building, we need to enrich the graphs with a lot more semantic information. Therefore it is necessary to classify indoor areas and to attach further class-dependent information to the nodes and edges. In this paper we illustrate the node types with two examples, namely 'rooms' and 'corridors'. Other types could be 'waiting area', 'meeting point' etc. In hierarchic graphs with many levels, node types like 'wing', 'storey', 'building' etc. are needed.

The node and edge types correspond directly to an ontology of building components. At present it is, however, not yet clear whether it is possible to describe the ontology in a formal description language like the Web Ontology Language, in short OWL[2], and to automatically incorporate the OWL concepts into the graph data structures. If this were indeed possible, it would make the graph framework much more elegant and flexible.

4.1   Rooms

Rooms which are not further decomposed into cells are represented by a single node. Each door is represented by an edge leading to the neighbouring rooms. This is not enough information for generating instructions like *"take the second door on your left"* with the optional clarification *"[the door] directly opposite the window to your right"*. In the event of further information being available, one could of course add the coordinates of the corners of a room and those of all

_____
[2] http://www.w3.org/TR/owl-features/

doors and windows. It turns out that for generating instructions like the ones above, it is sufficient to have a less complex data structure, such as a list of angles between the doors (or windows, respectively) and a reference line which goes from the centre of gravity of the room to a fixed reference point at the wall (we use the most north-western corner). An example is depicted in Fig. 4.

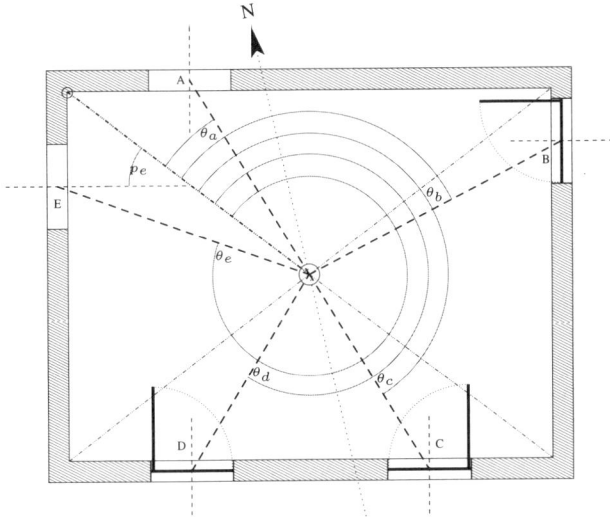

Fig. 4. Hybrid Model of a Room

A path crossing the room by entering through door B and leaving through door D may, for example, be described by the statement *"take the second door on your left"*. The information *"second door on your left"* can be computed as follows: the trajectory from B to the centre divides the room into *left* and *right*. Doors C and D are to the left and windows A and E are to the right. This can be derived from the angular distribution of the doors and windows. Thus, D is 'to your left'. The fact that D is the second door on your left can simply be obtained, by counting the number of doors in clockwise direction from B to D.

The further clarification *"[the door] directly opposite the window to your right"* can only be generated when the angular orientation of the walls is also stored. Together with the orientation of the doors and windows one can find out whether there exists another door or window which is situated opposite to door D.

The methods described above implicitly assume that the person entering at door B is looking towards the middle of the room. For this case it is sufficient to store a single angular distribution at the room node. If, however, the person looks straight forward when he enters the door, his notion of left and right may be different. Window E would now be to the left instead of right, for example. To account for this, one must compute the angular distribution for each door separately and store it at the corresponding end of the edge that represents the door. The line of reference for the angles crosses the middle of the door and is perpendicular to the door.

For many notions there are phrases in the human language which describe these notions with varying degree of precision. For example, there exist several degrees of *opposite*, such as *somewhat opposite*, *fairly opposite*, and *directly opposite*. A possible mathematical representation of these fuzzy notions are *fuzzy sets*, in our case fuzzy angular distributions (see Fig. 5)[3]: This has the advantage that deviations from an angle, like for the notion of being opposite, can still be regarded as being opposite, but only to a certain degree (determined by the membership, a fuzzy value between 0 and 1). The choice whether to use, for example, *somewhat opposite* or *fairly opposite* can be done by evaluating the corresponding fuzzy values on the distribution. If, say, the fuzzy value for the angle has been evaluated to 0.6, it would qualify as *somewhat opposite* whereas 0.95 would be considered as *directly opposite*. It is practical to use several intervals with decreasing threshold values for the various levels of 'opposing'.

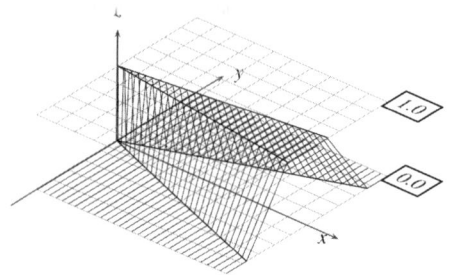

Fig. 5. Fuzzy Angular Distribution ($\theta_c > 0$, $\theta_s > \theta_c$)

4.2 Corridors

There are in fact two ways for modelling corridors. The first method is to decompose a corridor into cells such that each entrance to the corridor can be associated with a representative cell. Adjacent cells are represented by edges between the corresponding nodes. The main corridor in Fig. 1 can modelled this way, leading to a representation of seven cells for the seven adjoining doors. This representation is completely sufficient for solving wayfinding problems. It is, however, very cumbersome to generate a statement like *"take the third door to the right"* and from a practical point of view, it is certainly not the most elegant and compact[4] data structure.

The second way is illustrated in Fig. 6. The whole corridor is represented by a single node. However, this node actually stands for a directed linear structure leading from the front to the end, with openings both on its left hand side and its right hand side.[5] It does not affect the general notion of linearity whether the corridor is distorted, since the notions of 'left' and 'right' are relative and thus

---

[3] Depending on the application, core angles θ_c (fuzzy value of 1) and support angles θ_s (fuzzy value > 0) can vary.

[4] In terms of storage.

[5] In a way, the node is a dual to the linear structure.

change accordingly. The node must have labels which represent the entrances at the left side of the corridor, the entrances at the right hand side of the corridor, and the entrances at both ends of the corridor. The list of edges must reflect the real sequence of doors, stairways etc. It must keep the distances between two subsequent elements as well, or the offset from the front. Using these lists, it is easy to reconstruct from a particular door and a particular orientation an instruction like *"go to the second door on your left"*. The main corridor in Fig. 1 could be partitioned into several of such sequences.

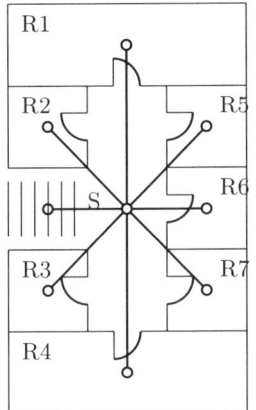

Labelling of the corridor node:
Left: R3,S,R2
Right: R7,R6,R
Front: R4
End: R1

Fig. 6. Corridor

5   Related Work

The modelling of indoor environments involves various research areas. In the studies of Franz et al. [3], aspects of both architecture and cognitive science are investigated. Most importantly, the versatility of graphs as formal models of built spaces is pointed out. Beyond occupancy grid models used in artificial intelligence, place graphs and view graphs are concepts contributed by cognitive science. Visibility plays a major role in defining where a place ends and another one starts. The *space syntax*[6] is a method which covers three elementary aspects of wayfinding: Access graphs, axial maps and isovist (or visibility) fields. In the field of robot navigation, *topological* and *cognitive maps* [7,2,13] are common practice for indoor models. The hybrid approach presented by Kuipers et al. [7] creates large-scale topological maps from small-scale metrical maps (with a local reference system). Similarly do Broch et al. [2] create global maps in a bottom-up fashion.

Hierarchies in buildings where entities can be entered and left through exit points on the boundary were discussed by Hu and Lee [5]. These entities can be clustered. Particularly, it is emphasised that for indoor wayfinding, topological

---

[6] http://www.spacesyntax.org/

relations of regions should reflect *reachability semantics* rather than mere intersection. Jiang et al. [6] introduce a model with location identifiers for hierarchies. Approaches from cognitive science pursue a functional perspective in which the intrinsic, egocentric viewpoint of a human wayfinder is adopted, as opposed to a bird's eye view offered e.g. by maps. In order to describe indoor environments with their inherent hierarchy by cognitive elements, *image schemata* [9,10] have been proposed. Examples include concepts like CONTAINER, REGION, GATEWAY and PATH, which are characterised by *affordances* (their specific function in social context), too.

Ontological aspects of indoor environments are covered e.g. by Bittner, Tsetsos et al. [1,12]. Bittner [1] proposes a formal characterisation of built environments by *partitions* which represent approximate regions and their boundary relations. Tsetsos et al. [12] focus their attention on application aspects and enhanced personalised indoor routing. A strong point in Bittner's approach is the clear, unambiguous definition of regions by boundaries. Furthermore, a distinction between so-called *bona-fide* (hard) and *fiat* (soft) *boundaries* is given: Hard boundaries are impenetrable, tangible barriers (like walls or fences). In contrast, soft boundaries are sometimes invisible, nonetheless existent barriers which are often unconsciously passed by the wayfinder (e.g. a marked area on the ground of a subterranean garage which delimits a particular car-park).

6   Conclusion and Future Work

We have introduced a novel model for indoor environments. Not only does it support wayfinding algorithms, but also it supports the generation of human-understandable path descriptions. For this purpose it combines quantitative with qualitative spatial information into a hybrid structure. We have shown how angular and distal expressions enrich the basic graph structure and enable reasoning within a local frame of reference. This will be needed for generating route instructions like *"take the 2nd door to the right"*. Some scenarios can require quantitative information, for instance, when querying the total number of seats in a classroom or its capacity for storing goods.

The graphs presented in our model are hierarchical. Graphs at a lower level are refinements of nodes (or edges) at a higher level. The nodes and edges are of different types and carry different kinds of information, which is not solely the physical distance. The types reflect an ontology of rooms, buildings, and so on.

The graphs together with some of the standard algorithms, shortest paths, nearest neighbours etc., have been implemented in the prototypic TransRoute system [11]. In a next step we want to incorporate context information into the wayfinding algorithms. The context information can come from the environment itself (e.g. a door is closed at night), or it can date from user models (e.g. a preference for lifts instead of stairs).

For generating descriptions of paths, we are currently developing a markup language. The elements of this language describe operations for navigating indoor and outdoor environments.

Acknowledgements

This research has been co-funded by the European Commission and by the Swiss Federal Office for Education and Science within the 6th Framework Programme project REWERSE number 506779 (cf. http://rewerse.net).

References

1. Thomas Bittner. The Qualitative Structure of Built Environments. In *Fundamenta Informaticae*, volume 46 (nr. 1–2), pages 97–128, 2001.
2. Juan Carlos Peris Broch and Mara Teresa Escrig Monferrer. Cognitive Maps for Mobile Robot Navigation: A Hybrid Representation Using Reference Systems. In Michael Hofbaur, Bernhard Rinner, and Franz Wotawa, editors, *Proceedings of the 19th International Workshop on Qualitative Reasoning (QR-05)*, pages 179–186, 2005.
3. Gerald Franz, Hanspeter Mallot, and Jan Wiener. Graph-based Models of Space in Architecture and Cognitive Science - a Comparative Analysis. In Y-T. Leong and G.E. Lasker, editors, *Proceedings of the 17th International Conference on Systems Research, Informatics and Cybernetics (InterSymp'2005)*, Architecture, Engineering and Construction of Built Environments, pages 30–38. The International Institute for Advanced Studies in System Research and Cybernetics, 2005.
4. Christoph Hölscher, Tobias Meilinger, Georg Vrachliotis, and M. Knauff. The Floor Strategy: Wayfinding Cognition in a Multi-Level Building. In Akkelies van Nes, editor, *Proceedings of the 5th International Space Syntax Symposium*, volume 2, pages 823–824. Techne Press, June 2005.
5. Haibo Hu and Dik Lun Lee. Semantic Location Modeling for Location Navigation in Mobile Environment. In *Proceedings of the 5th IEEE International Conference on Mobile Data Management (MDM 2004)*, pages 52–61. IEEE Computer Society, 2004.
6. Changhao Jiang and Peter Steenkiste. A Hybrid Location Model with a Computable Location Identifier for Ubiquitous Computing. In Gaetano Borriello and Lars Erik Holmquist, editors, *Proceedings of the 4th International Conference on Ubiquitous Computing (UbiComp 2002)*, volume 2498 of *Lecture Notes in Computer Science*, pages 246–263. Springer, 2002.
7. Benjamin Kuipers, Joseph Modayil, Patrick Beeson, Matt MacMahon, and Francesco Savelli. Local metrical and global topological maps in the Hybrid Spatial Semantic Hierarchy. In *Proceedings of the 2004 IEEE International Conference on Robotics and Automation (ICRA 2004)*, volume 5 of *ICRA*, pages 4845–4851. IEEE Computer Society, 2004.
8. Kevin Lynch. *The Image of the City*. MIT Press, June 1960.
9. Martin Raubal and Michael Worboys. A Formal Model of the Process of Wayfinding in Built Environments. In Christian Freksa and David M. Mark, editors, *Spatial Information Theory: Cognitive and Computational Foundations of Geographic Information Science, Proceedings of the International Conference COSIT '99*, volume 1661 of *Lecture Notes in Computer Science*, pages 381–401. Springer, 1999.
10. Urs-Jakob Rüetschi and Sabine Timpf. Modelling Wayfinding in Public Transport: Network Space and Scene Space. In Christian Freksa et al., editor, *Spatial Cognition IV*, volume 3343 of *Lecture Notes in Artificial Intelligence*, pages 24–41. Springer, 2005.

11. Edgar-Philipp Stoffel. A Research Framework for Graph Theory in Routing Applications. Diplomarbeit/diploma thesis, Institute of Computer Science, LMU, Munich, 2005. http://www.pms.ifi.lmu.de/publikationen/#DA_Edgar.Stoffel.
12. Vassileios Tsetsos, Christos Anagnostopoulos, Panayiotis Kikiras, Tilemahos Hasiotis, and Stathes Hadjiefthymiades. A Human-centered Semantic Navigation System for Indoor Environments. In *Proceedings of the IEEE/ACS International Conference on Pervasive Services (ICPS'05)*, pages 146–155. IEEE Computer Society, 2005.
13. Jan Oliver Wallgrün. Hierarchical Voronoi-based Route Graph Representations for Planning, Spatial Reasoning, and Communication. In Patrick Doherty, editor, *Proceedings of the 4th International Cognitive Robotics Workshop (CogRob-2004)*, pages 64–69, 2004.

MEMS Mobile GIS: A Spatially Enabled Fish Habitat Management System

A. Rizzini[1], K. Gardiner[2], M. Bertolotto[1], and J. Carswell[2]

[1] School of Computer Science and Informatics, University College Dublin, Belfield,
Dublin 4, Ireland
[2] Digital Media Centre, Dublin Institute of Technology, Aungier Street, Dublin 2,
Ireland

Abstract. Spatially enabled computing can provide assistance to both web-based and mobile users by exploiting positional information and associated contextual knowledge. The Mobile Environmental Management System (MEMS) is a proof of concept prototype that has been developed in order to simplify administrative duties of biologists at the Department of Fisheries and Oceans (DFO), Canada. MEMS aims to deliver context-aware functionality aided by visualization, analysis and manipulation of spatial and attribute datasets. The resulting application delivers a set of functions and services that aids the DFO's biologists in making everyday management decisions.

1 Introduction

This paper describes the techniques used in the development of a web-based and Mobile Environmental Management System (MEMS). The current prototype has been tailored to deliver context-aware functionality aided by visualization, analysis and manipulation of spatial and attribute datasets. Context associated knowledge is an intelligent process which retrieves specific data for users. It is achieved by combining knowledge gained about data processed in the past with the activities planned by the user, together with other activity dependencies such as geographical location. The MEMS datasets are provided by the Canadian Department of Fisheries and Oceans (DFO) and the prototype is customized to the specific needs of the Great Lakes Laboratory for Fisheries and Aquatic Sciences (GLLFAS) Fish Habitat Management Group's requirements for fish species at risk assessment. Currently, biologists have only access to the fisheries data from their office. This greatly prevents them from interacting with the data in a real-time environment, reducing their productivity and effectiveness in the field. Spatially enabling a mobile device allows mobile GLLFAS biologists to make informed decision immediately. This research concerns DFO priorities specifically to administer the fish habitat provision of the Fisheries Act, in particular those that are aimed at preventing the harmful alteration, disruption or destruction of fish habitat. This is done to conserve, restore and develop the productive capacity of habitats for recreational, commercial and subsistence fisheries both in the freshwater and marine environments[1, 2]. The functionality that GLLFAS

J.D. Carswell and T. Tezuka (Eds.): W2GIS 2006, LNCS 4295, pp. 113–122, 2006.
© Springer-Verlag Berlin Heidelberg 2006

biologists require from the MEMS prototype includes access to geo-referenced maps and imagery, to overlay the current position on a map and to manipulate (e.g. input/edit/query) attribute data in the field while wirelessly connected (where possible) to the office database[12]. Additional functionality also required by the DFO is the ability to record, edit and view multimedia annotations, perform scientific/common-name conversion and graph generations of results. The current "fish species at risk" work-flow, whereby scientists enter textual/ pictorial information on paper field data sheets is inefficient, has potential for inaccuracies during both initial recording and subsequent data entry phases, and does not facilitate knowledge sharing between staff. Also, different types of information may be stored in different locations and valuable time can often be lost trying to correlate data in order to make decisions. The proposed MEMS system has the following advantages over current practice:

1. Facilitates knowledge sharing and data analysis/synthesis.
2. Supports effective communication between different staff at different physical locations (e.g. scientists in the lab and colleagues in the field).
3. Allows important multimedia data and associated annotations to be combined with text-based records.
4. Saves time and money by reducing paperwork and allows staff to input and access information anywhere at any time without having to return to dedicated access points.
5. Reduces error by reducing latency between collection and data entry, as well as paperwork.

2   Related Work

There are few contemporary systems that can be compared to the current prototype. One in particular is the Mobile Environment Monitoring System (MEMoS)[3] developed at the Multi-purpose Environmental Modelling Facility of the University of Windsor in Canada in collaboration with the Conestoga-Rovers & Associates (CRA). This system is designed to deliver real-time environmental information using rugged field computers, radio technologies (CRA's OpenRTU [4]) and highly modular, interoperable software. The system also allows the user to rapidly collect and integrate spatially referenced data from a range of mobile environmental data collectors and update a GIS data warehouse in real-time making data available to a web-based client application instantly. This system adopts a distributed approach and uses a remote base station and mobile data loggers. The remote base station is a vehicle, containing the OpenRTU receiver and a computer with an Oracle Spatial database server (see Figure 1). The OpenRTU system is a remote process monitor that collects data from connected data loggers. The station allows multiple roving data collectors to wirelessly send recorded structures, where the incoming data are routed to appropriate tables in a local database. The desktop, in the base station, contains a web-enabled client application which is used to generate real-time results for the data collected. The mobile data loggers are backpack type equipment containing the OpenRTU transmitter connected to a series of environmental sensors and a rugged Tablet PC for data input.

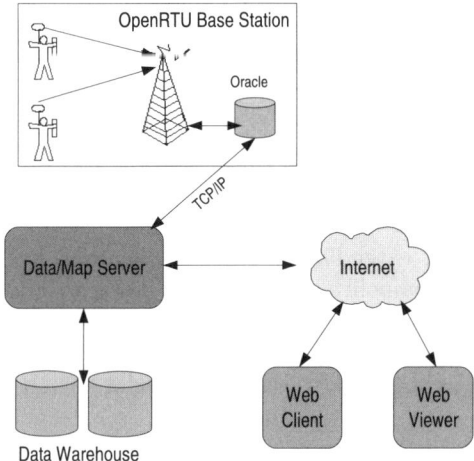

Fig. 1. MEMoS System Architecture

When the base station has finished collecting data, it connects via wired or wireless TCP-IP to the data warehouse, updating the new datasets. The advantage of this system is that it uses an inexpensive communication infrastructure between the base station and the client station. If the communication had to be established via GPRS, running cost could dramatically escalate. A possible disadvantage in running the OpenRTU transmission medium, is the actual working range. Also, the disjunctive update of the data warehouse means that information may not be accurate until such updates are completed.

3 Technologies

The MEMS prototype uses an Oracle Spatial database. Oracle Spatial provides a platform that supports a wide range of applications from automated mapping/facilities management and Geographic Information Systems (GIS), to wireless location services and location-enabled e-business. Oracle Spatial is integrated into the extensible Object Relational Database Management System (ORDBMS), which allows access to the full functionality and security of the underlying DBMS[7]. Along with the database, the Oracle proprietary OC4J application server is used. The application server is a component of Oracle, and is installed automatically. The application server acts as deployment platform for Oracle applications. The Oracle Enterprise Manager application is installed in the OC4J as a web application where the user can manage the database. In addition MEMS uses eSpatial Solutions[8] iSmart Suite to consolidate all forms of spatial data. The suite is based on a J2EE application development environment where the user can easily integrate spatial components. iSmart is a collection of tools that enables developers to build and deploy spatial applications using a set of standard procedures. These tools offer developers a high-level development environment which is several time faster than developing the application from a base Java Development Environment.

4   System Architecture

The MEMS prototype uses a typical three-tier architecture[5] (see Figure 2) for enterprise information systems, composed of the client layer, application server layer, and the database layer. This architecture focuses on the development of services for a versatile, extendible (J2EE) application, instead of giving GIS capabilities to a large monolithic application. The implementation of the application follows a modularization approach. Functionality is encapsulated in packages avoiding cross functionality dependencies and hence enforcing stability. The communication between the client layer and the database are conducted

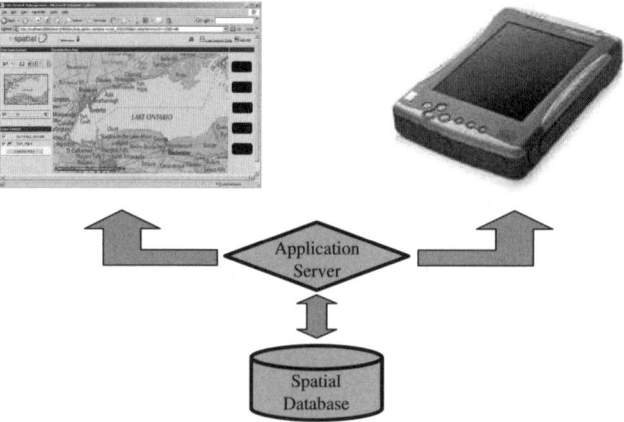

Fig. 2. MEMS System Architecture

through the application server layer. The application server layer acts as a mediator. It ensures a minimum amount of data is transferred between the server and the client. Requests from the client are interpreted and executed by the application server. A response is assembled and sent back to the client. With this type of architecture, the processing load is balanced, requiring a minimum level of computing power on a thin client[11]. The client layer consists of a comprehensive web-interface that provides biologists with the ability to input, edit, analyze and annotate data over the Internet. A Tablet PC[6] is used equipped with a Global Positioning System (GPS) receiver and a General Packet Radio Service (GPRS) network connection. As the user navigates in the field their positioning and orientation is displayed on the web-interface on a geo-referenced map.

5   System Functionality and GUI

The MEMS prototype uses a comprehensive graphical user interface (GUI) to interact with the user. The GUI is divided into three main parts. The left panel (See Figure 3) is the "Navigation panel". It contains navigation functions used by the user to navigate the map (zoom to area, click x/y coordinates, zoom to extent, back to original view, previous view, refresh map). The navigation map

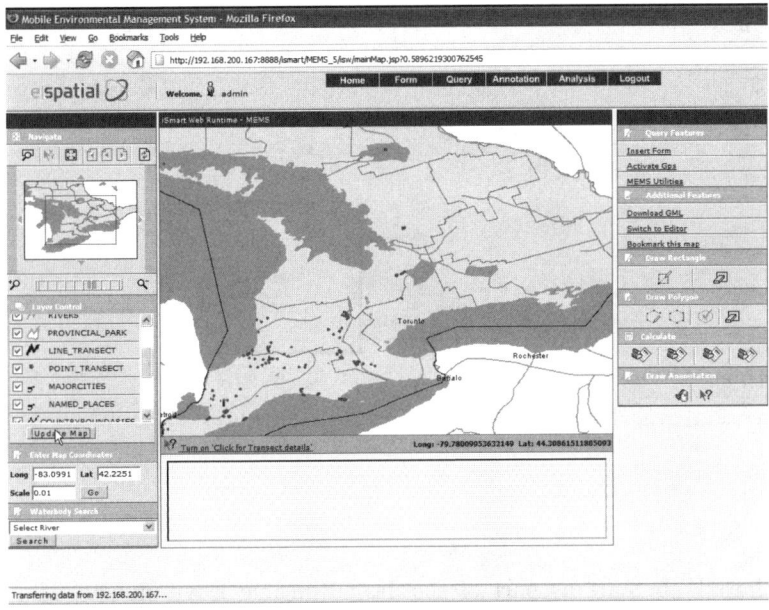

Fig. 3. MEMS Graphical User Interface

contains a red square which represents the area displayed by the main map. This is designed to help the user navigate more effectively when zoomed into a small area on a map. The "Layer Control" is located below the navigation map and is used to select layers on the main map. The last two controls on the navigation panel contain "Map Coordinates", used to centre the main map with any given coordinates and the "Search River" function control, used for searching recorded habitats. In the centre of the application (See Figure 3) the "Map Pannel" contains the main map. The main map offers "Clickable Map" functionality that enables the user to select features on the map and displays information about them. The "Tools panel" contains controls for functions which are explained in the following sections.

5.1   GPS Acquisition

This is a core module as the distributed MEMS application forces the client to acquire GPS coordinates first and then send them to the application server.

As the client was not designed to be installation independent it is necessary to execute the Java application remotely. One of the technologies that Java offers is JNLP (Java Network Launching Protocol). JNLP applications are launched

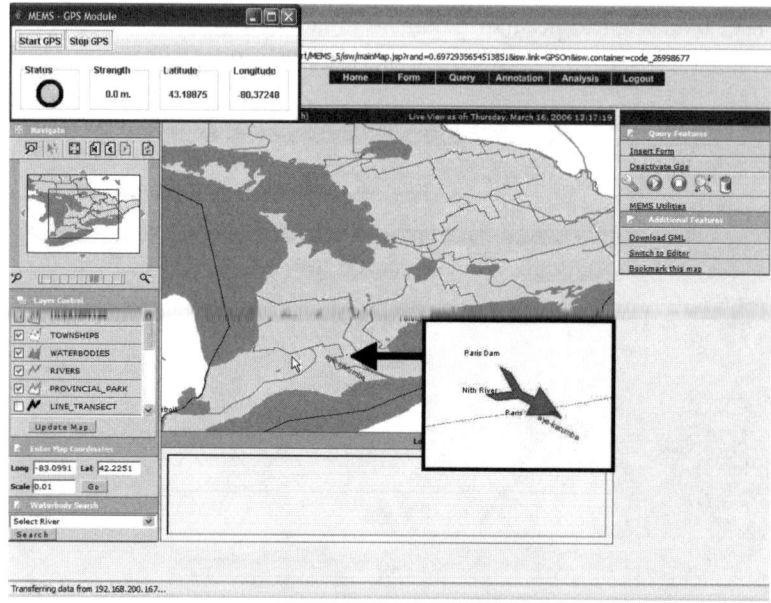

Fig. 4. Gps Module

using Java Web Start as part of the Java Runtime Environment. The corresponding GPS application registers each client IP address and host-name with the application server so that multiple feeds can be displayed on the map. The screenshot (See Figure 4) shows the GPS acquisition overlayed on the MEMS map. In the top left corner the GPS module control is shown and is launched by the java Web Start while in the center the black square is a zoomed image of the GPS feed.

5.2   Multimedia Annotation

Multimedia annotation is an advanced feature of MEMS. Annotation is a simple way to record data on the fly. For example, a new species of fish could be encountered and visual evidence would be of great assistance. This functionality enables the user to embed video, audio, text and image annotations on the map. These annotations are uploaded to the database as BLOB data along their associated coordinates. Using JSP and Servlet technology, the user is required to enter some text describing the annotation. If the annotations are video, audio or image annotations a file is also required to be uploaded. The file byte-stream is transferred to the application server, along the text and the current location, where it is inserted in the database. When this procedure is completed the map

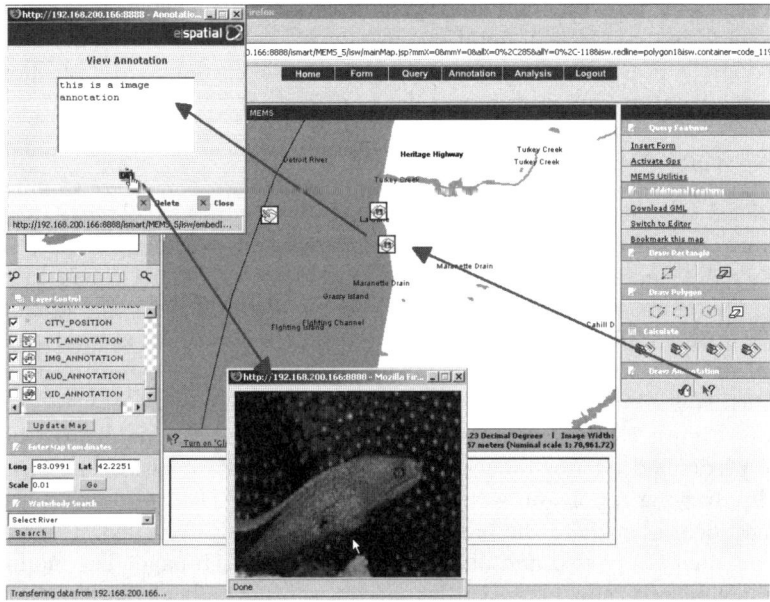

Fig. 5. Multimedia Annotation

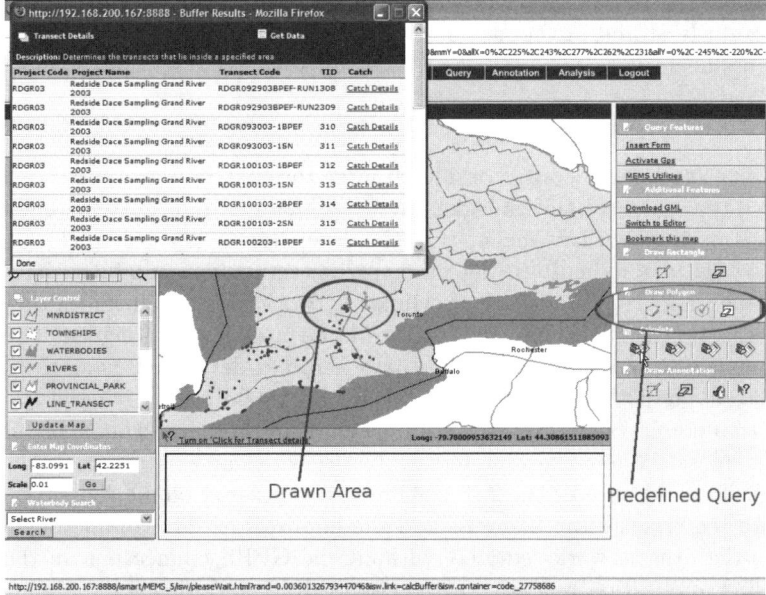

Fig. 6. Spatial Queries

is refreshed and an icon representing the annotation is displayed. (See Figure 5). The user can view the annotations by clicking on a corresponding icon. This action opens a pop-up window that displays any text associated with the annotation and a link to any multimedia data.

5.3   Spatial Queries

The Spatial Queries Tool is an advanced tool which enables biologists to quickly and easily query the database. Spatial Queries are complex in nature and require a great deal in understanding of Oracle Spatial and the Sequel Query Language (SQL). Therefore, an easy to uses interface was developed to enable biologists to perform them. The tool is very easy to use. The biologists are only required to draw an area on the map and it is then possible to execute a number of predefined spatial queries on the selected area. The polygon highlighted on the map (See Figure 7)is the actual spatial component and the highlighted buttons are the predefined queries requested by the GLLFAS biologists. The procedure starts by drawing a polygon or a square on the map. The red dots on the map represent locations where the biologists have previously recorded. The result of any given query is parsed and displayed in a pop-up web-page. The module has been implemented using extensive JavaScript to draw and calculate the points of the polygon. Once the polygon is drawn a JSP is used to transform the polygon to an Oracle JGeometry object. This object is then stored in the user session ready for use in the query. Using iSmart, a SQL command is embedded in the application and is executed when the query button is clicked upon. The result is temporarily stored in the user session, where it will be parsed and displayed by the application.

5.4   Offline Module

The offline module was only considered after the first visit to the DFO. It was observed that the area the biologists were working had intermittent cellular signal. This unexpected fact meant that the development of a backup system was necessary. It was difficult to detect or predict network availability but fortunately the biologist were aware of cellular network presence in areas they need to investigate. Using this information the offline module was designed. The module is implemented using stand-alone Java and it is developed using the iSmart technology. The Offline Module also called MEMSOffline is a standalone application that connects directly to the online application server where the biologist is asked to select an area of interest. The selected area is compressed into a zip type file and saved. The MEMSOffline compressed offline file contains all the data required by the biologist and the application. This offline application enables the user to work remotely without the GPRS connection. In the field, the offline application reads in the compressed file and displays the map using the iSmart Editor application, enabling the biologists to perform spatial queries, insert new forms, look at previously recorded forms, display GPS feeds and navigate the map. It also offers restricted multimedia annotation functionality. Data which is changed or added during the sampling session is stored and saved in

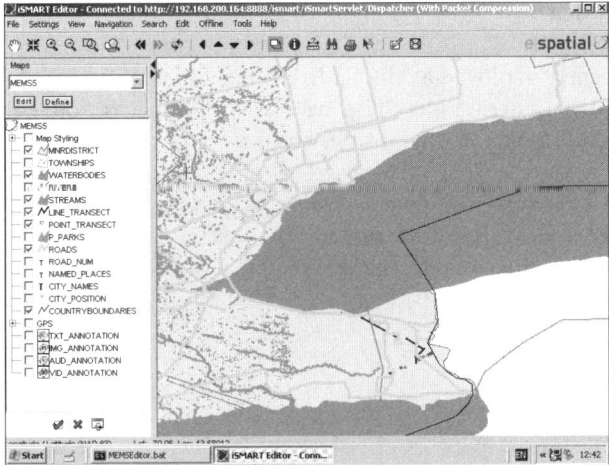

Fig. 7. MEMSOffline

the offline file. When the application establishes a network connection and can connect to the application server, the data is synchronized with the server and any additions are added to the database.

5.5 Forms Module

This Forms module was requested from the GLLFAS biologists and is designed accordingly. A requirement for this part is that the biologists need to navigate back and forward from different "fish forms", without committing them to the database, until all the data has been successfully collected. This behavior proved difficult to implement in a Client/Server environment. In order to achieve this AJAX (Asynchronous JavaScript and XML) technology is used. AJAX[9] uses a series of JavaScripts to compose and send a URL request to a server side page (JSP, Servlet, PHP, CGI) and return and display the response (using DOM objects) in the web-application. This is done without the need to reload the page. Using this technology, the form changes can be posted to the corresponding Java data object in the session. If the user changes a field in the form, the corresponding Java object is updated immediately. This module also offers the ability to dynamically add new "fish forms" as required. If biologists encounter a species which has not been recorded they can dynamically add another field to the current form to accommodate it. This module has also been designed to easily handle database errors or warnings that may arise during program execution. If constraints are triggered, they can easily be interpreted and displayed the user. If the unique project code has been previously entered in the database a warning message is displayed to the user. Another feature is the ability to validate the fields in the forms. Extensive JavaScripts are used to iterate through the fields of the page and check for types (Double-Integers), range and white spaces.

6 Conclusions

MEMS is currently deployed at the GLLFAS headquarters, where extensive field-ing testing is been carried out. The system is fully operational and it offers a spatially-enabled mobile and adaptable service for the biologists, ensuring better utilization of resources. The advantages of the system go beyond the system's functionality. This system delivers services which improve the biologists work-ing environment. On going work include the testing phase of the system. In the future we intend to investigate the adaptability of the architecture and GUI to different applications.

References

[1] Minns, C.K. "Quantifying 'No Net Loss' of Productivity of Fish Habitats", Cana-dian Journal of Fisheries and Aquatic Sciences (Journal canadien des sciences halieutiques et aquatiques), 54 , pp. 2463-2473, 1997.

[2] Minns, C. K., "Science for Freshwater Fish Habitat Management in Canada: Cur-rent Status and Future Prospects". Aquatic Ecosystem Health and Management, 4, pp. 423-436, 2001.

[3] Graniero, P.A. and Miller, H.S. "A mobile environmental monitoring system with real-time database updates." CRESTech Innovation Network Annual Meeting, Toronto ON, 2002.

[4] OpenRTU, eSolutionsGroup, http://www.openrtu.com/

[5] Pierce, M., Youn, C.-H., Fox, J., "The Gateway Computational Web Portal: Devel-oping Web Services for High Performance Computing", International Conference on Computational Science, 1, pp. 503-512, 2002.

[6] Hp TR3000, http://h18000.www1.hp.com/products/quickspecs/11909_na.PDF

[7] Sharma, J., "Oracle Spatial", An Oracle technical white paper, Oracle Corp., Red-wood City, CA, May 2001.

[8] eSpatial Solutions, http://www.espatial.com/

[9] Olson, S., "Ajax on Java", O'Reilly Media, 2006, ISBN: 0596101872

[10] Marinilli, M., Java Deployment Using JNLP and WebStart, September 2001, ISBN: 0672321823

[11] Carswell, J. D., Gardiner K., Bertolotto M., Mandrak N., "Applications of Mobile Computing for Fish Species at Risk Management" Proceedings of International Conference on Environmental Informatics of International Society of Environmen-tal Information Sciences (ISEIS2004), Regina, Canada, 2004.

[12] Gardiner, K., Rizzini, A., Carswell, J., Bertolotto, M. "MEMS: Mobile Environ-mental Management System", GIS Research UK, 13th Annual Conference, Uni-versity of Glasgow, 6th-8th April 2005.

A Mobile Computing Approach for Navigation Purposes

Mohammad R. Malek[1,2] and Andrew U. Frank[3]

[1] Dept. of GIS, Faculty of Geodesy and Geomatics Eng., KN Toosi University of Technology,
Tehran, Iran
[2] Dept. of Surveying and Geomatics Eng., University of Tehran, Tehran, Iran
[3] Institute for Geoinformation and Cartography, Technical University Vienna, Austria
malek@ncc.neda.net.ir, frank@geoinfo.tuwien.ac.at

Abstract. The mobile computing technology has been rapidly increased in the past decade; however there still exist some important constraints which complicate the use of mobile information systems. The limited resources on the mobile computing would restrict some features that are available on the traditional computing technology. In almost all previous works it is assumed that the moving object cruises within a fixed altitude layer, with a fixed target point, and its velocity is predefined. In addition, accessibility to up-to-date knowledge of the whole mobile users and a global time frame are prerequisite. The lack of two last conditions in a mobile environment is our assumptions. In this article we suggest an idea based on space and time partitioning in order to provide a paradigm that treats moving objects in mobile GIS environment. A method for finding collision-free path based on the divide and conquer idea is proposed. The method is, to divide space-time into small parts and solve the problems recursively and the combination of the solutions solves the original problem. We concentrate here on finding a near optimal collision-free path because of its importance in robot motion planning, intelligent transportation system (ITS), and any mobile autonomous navigation system.

Keywords: Mobile GIS, Mobile computing, Navigation, Free-collision path, Optimization.

1 Introduction

Mobile agents and movement systems have been rapidly increased worldwide. Within the last few years, we were facing many advances in wireless communication, computer networks, location-based engines, and on-board positioning sensors. Mobile GIS as an integrating system of mobile agent, wireless network, and some GIS capabilities has fostered a great interest in the GIS field [13]. Without any doubt navigation and routing could be one of the most popular GIS based solution on mobile terminals. Due to this fact the mobile GIS is defined as an area about non-geographic moving object in geographic space [19].

Although the mobile computing has been increasingly growing in the past decade, there still exist some important constraints which complicate the use of mobile GIS systems. The limited resources on the mobile computing would restrict some features that are available on the traditional computing. The resources include computational

J.D. Carswell and T. Tezuka (Eds.): W2GIS 2006, LNCS 4295, pp. 123–134, 2006.
© Springer-Verlag Berlin Heidelberg 2006

resources (e.g., processor speed and memory) user interfaces (e.g., display and pointing device), bandwidth of mobile connectivity, and energy source [2], [10], [19], and [32]. In addition, one important characteristic of such environment is frequent disconnection that is ranging from a complete to weak disconnection [10] and [38]. The traditional GIS computation methods and algorithms are not well suited for such environment. These special characteristics of mobile GIS environment make us pay more attention to this topic.

In this paper, finding a path without any conflict which is so-called collision-free path is highlighted. It is an important task of routing and navigation. Collision-free path and its variants find applications in robot motion planning, intelligent transportation system (ITS), and any mobile autonomous navigation system. It will be concluded that Wayfinding which is a fundamental spatial activity that people experience in daily lives, could be solved by this method.

Within the framework of this paper we attempt to apply an idea to treat moving objects in mobile GIS environment based on partitioning in space and time. The idea is, to divide space-time into small parts and find solution (e.g. collision-free paths and wayfinding procedure) recursively. The connecting results will be the collision-free path. This paper addresses the problem of finding collision-free path in mobile GIS environment. The rest of this paper is organized as follows: Section 2 reviews the related works. Section 3 describes our suggested methodology. Then, section 4 explains how one can find a collision-free path by an optimization problem. Finally, we give concluding remarks.

2   Related Works

An overview of collision detection, mathematical methods, and programming techniques to find collision-free and optimal path between two states for a single vehicle or a group of vehicles can be found in [7], [16], and [34], respectively. In the field of robot motion planning potential field methods introduced by Khatib, are widely used [17]. The main attraction of potential method is its ability to speed up the optimization procedure. Path planning techniques using mixed-integer linear program were developed earlier, especially in the field of aerial vehicles navigation (see e.g. [27-28], [29-30], and [33]. The reader who wants to see more related topics is referred to [11]. In almost all works it is assumed that the moving object cruises within a fixed altitude layer, with a fixed target point, and its velocity is predefined. In addition, accessibility to up-to-date knowledge of the whole mobile agents and a global time frame are prerequisite. The lack of two last conditions in distributed mobile computing environment is a well-known fact.

A method for reducing the size of computation is computation slice [12] and [25]. The computation slicing as an extension of program slicing is useful to narrow the size of the program. It can be used as a tool in program debugging, testing, and software maintenance. Unlike a partitioning in space and time, which always exists, a distributed computation slice may not always exist [12].

Among others, two works using divide and conquer idea, called honeycomb and space-time grid, are closer to our proposal. The honeycomb model [8] focuses on temporal evolution of subdivisions of the map, called spatial partitions, and gives a formal semantics for them. This model develops to deal with map and temporal map only. The concept of space-time grid is introduced by Chon et al. [4-6]. Based upon the space-time grid, they developed a system to manage dynamically changing information. In the last work, they attempt to use the partitioning approach instead of an indexing one. This method can be used for storing and retrieving the future location of moving object.

In the previous work of the first author [20-23] a theoretical framework using Influenceability and a qualitative geometry in the mobile environment with application in the relief management was presented. This article can be considered as an empirical extension of them.

3   Algebraic and Topological Structure

Causality is a well-known concept. There is much literature on causality, extending philosophy, physics, artificial intelligence, cognitive science and so on (e.g. [1, 14, 34]). In our view, influenceability stands for spatial causal relation, i.e. objects must come in contact with one another; cf. [1]. Although influenceability as a primary relation does not need to prove, it has some exclusive properties which show why it is selected. Influenceability supports contextual information and can be served as a basis for context aware mobile computing which has attracted researchers in recent years [9] and [26]. This relation can play the role of any kind of accident and collision. It is well-known that the accident is the key parameter in most transportation systems (for example see [31]). As an example the probability of collision defines the GPS navigation integrity requirement. In addition, this model due to considering causal relation is closer to a naïve theory of motion [25].

In the relativistic physics [15] based on the postulate that the vacuum velocity of light c is constant and maximum velocity, the light cone can be defined as a portion of space-time containing all locations which light signals could reach from a particular location (Figure 1). With respect to a given event, its light cone separates space-time into three parts, inside and on the future light cone, inside and on the past light cone, and elsewhere. An event A can influence (influenced by) another event; B; only when B (A) lies in the light cone of A (B). In a similar way, the aforementioned model can be applied for moving objects. Henceforth, a cone is describing an agent in mobile GIS environment for a fixed time interval. That means, a moving object is defined by a well-known acute cone model in space-time. This cone is formed of all possible locations that an individual could feasibly pass through or visit. The current location or apex vertex and speed of object is reported by navigational system or by prediction. The hyper surface of the cone becomes a base model for spatio-temporal relationships, and therefore enables analysis and further calculations in space-time. It also indicates fundamental topological and metric properties of space-time.

As described in Malek [21, 23], the movement modeling, are expressed in differential equation defined over a 4-dimensional space-time continuum. The assumption of a 4-dimensional continuum implies the existence of 4-dimensional spatio-temporal parts. It is assumable to consider a continuous movement on a differential manifold M which represents such parts in space and time. That means every point of it has a neighborhood homeomorphic to an open set in R^n. A path through M is the image of a continuous map from a real interval into M. The homeomorphism at each point of M

determines a Cartesian coordinate system (x_0, x_1, x_2, x_3) over the neighborhood. The coordinate x_0 is called time. In addition, we assume that the manifold M can be covered by a finite union of neighborhoods. Generally speaking, this axiom gives ability to extend coordinate system to the larger area. This area shall interpret as one cell or portion of space-time. The partitioning method is application dependent. The partitioning method depends on application purposes [5] on the one hand, and limitation of the processor speed, storage capacity, bandwidth, and size of display screen on the other hand. It is important to note that the small portion of space and time in this idea is different from the geographical area covered by a Mobile Supported Station (MSS). This idea is similar to Helmert blocking in the least squares adjustment calculation [36].

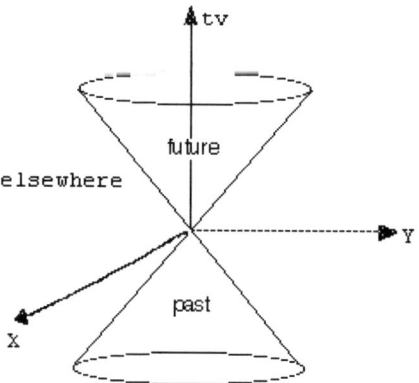

Fig. 1. A cone separates space-time into 3 zones, past, future, and elsewhere

Let us take influenceability as an order relation (symbolized by $\prec$) be primitive relation. It is natural to postulate that influenceability is irreflexive, antisymmetric, but transitive, i.e.,

$$(x \prec y) \wedge (y \prec z) \Rightarrow x \prec z \qquad (1)$$

Thus, it can play the role of 'after'.

Definition 1 (Temporal order): Let x and y be two moving objects with t_x and t_y corresponding temporal orders, respectively. Then,

$$(x \prec y) \Rightarrow (t_x < t_y) \qquad (2)$$

Connection as a reflexive and symmetric relation [10]can be defined by influenceability as follows:

Definition 2 (Connect relation): Two moving objects x and y are connected if the following equation holds;

$$(\forall xy)C(x, y) := [(x \prec y) \vee (y \prec x)] \wedge \{\neg(\exists a)[(x \prec a \prec y) \vee (y \prec a \prec x)]\} \quad (3)$$

Consequently, all other exhaustive and pairwise disjoint relations in region connected calculus (RCC) [3], i.e., *disconnection* (DC), *proper part* (PP), *externally connection* (EC), *identity* (EQ), *partially overlap* (PO), *tangential proper part* (TPP), *nontangential proper part* (NTPP), and the inverses of the last two; TPPi and NTPPi; can be defined.

The consensus task as an acceptance of the unique framework in mobile network can not be solved in a completely asynchronous system, but as indicated by Malek [21] with the help of influenceability and partitioning concept, it can be solved. Another task in mobile network is leader election. The leader, say a, can be elected by the following conditions:

$$\forall x \in \{The\ set\ of\ moving\ objects\} : a \prec x.$$

Furthermore, some other relations can be defined, such as which termed as *speed-connection* (SC) and *time proper overlap* (TPO) (see Figure 2):

$$SC(x,y) := \neg EQ(x,y) \wedge$$
$$\{[C(x,y) \wedge (\forall ab)(EC(x,a) \wedge (EC(x,b) \wedge EC(y,a) \wedge EC(y,b)] \Rightarrow C(a,b)\} \quad (4)$$

$$TPO(x, y) := \{(x \prec y) \wedge (PO(x, y) \wedge [\forall z\ (SC(x, z) \Rightarrow PO(y, z))]\}$$

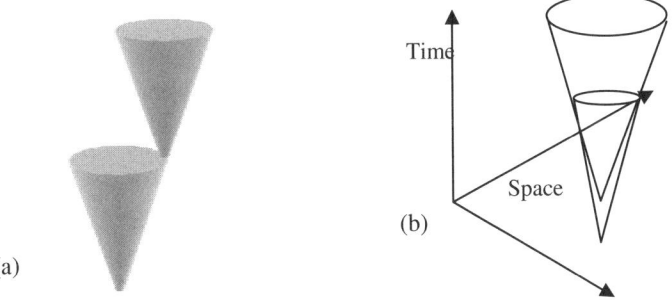

(a)   (b)

Fig. 2. a) Speed-connection relation and **b)** Time-proper relation between two objects

4   Collision-Free Path

An important task in navigation systems is to find a secure or collision-free path. A collision-free path is a route that a moving object does not have any collision or intersection with obstacles as well as other moving objects. It will be distinguished between two network architectures, centralized and co-operative. In the centralized architecture, a control center exists which receives and sends data to moving objects. In the co-operative architecture all moving objects exchange information between themselves [18]. In the former architecture, the control variables of all nodes associates in the optimization, but in the later only variables of the active node are considered. Finding a collision-free path requires four steps, dividing the space domain into small parts, finding connected cones, computing free space, and finally solving an optimization problem. The problem discussed in this section is using a mathematical programming technique to find the optimal or near optimal collision-free path between moving objects. The details of the other steps are left for future articles.

A mobile terminal in the mobile GIS environment exacerbates the tension between two extreme points. On the one hand, the resource poverty leads to the client-server architecture that the mobile host only supports a user interface but no application (dump terminal). On the other hand, frequent disconnections (see Figure 3), power saving, and scalability concerns of the server suggest a relatively high independent mobile host. Hence, the need to balance between these extreme roles of mobile host is necessary.

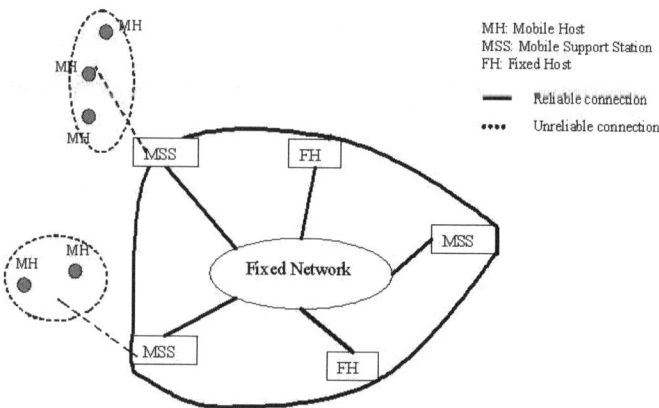

Fig. 3. In the mobile environment the fixed host communicates with a mobile host through a wireless interface that is provided by a mobile supported station

Let us continue with the following scenario: A private company in order to attract more tourists to the lake "Wörthersee" in Austria provides an autonomous navigation system for their motorized small boats. Each boat equipped with a palm-top computer; using GPS for positioning; that can communicate via a wireless network. Based upon this capability, the system can play the role of online tourist guide at each part of the lake. The server sends the necessary information like current position and velocity of the other boats those are relevant to the mobile host. That means, the only information of the agents that have accident possibility in the current cell will be sent. As an instance, only the information of the agent number 2 will be sent to the agent number 1 in the Figure 2. This rule can be formalized by spatial influenceability relation [20]. The mobile host will send its state information like its position and velocity once they have significant changes.

At call setup an optimal route is generated. This task can be done by the fixed host. In each cell, the preliminary route is ensured that no collision occurred. It is natural to define the target function by minimizing the distance between calculated route and the optimal one. It may be named as nearest to optimal path. The method described in this part provides a minimum distance formulation (1). It is combined with the linear collision avoidance constraints [33], turn and velocity constraints, and is extended to match with partition and conquer idea.

$$
\textit{Min. } \mathbf{d}^{\mathrm{T}} \, \Sigma \, \mathbf{d}
$$

S.T.:

Collision avoidance, turn, and velocity conditions

(1)

$\mathbf{d}$ is the vector of distances between optimal state parameters and the estimated control parameters in the space-time grid of interest. Finally, the linear constraint quadratic optimization problem should be solved. This part can be run in the clients and the procedure will repeat in other parts.

4.1 Collision Avoidance Condition

We shall consider for simplicity of exposition of two moving objects in a two-dimensional space. The position of agent p at time step i is given by (x_p^i, y_p^i) and its velocity by (v_{xp}^i, v_{yp}^i), forming the elements of the state vector $\mathbf{S}_{xp}^i$. The real value of the state parameter is represented by an asterisk. At every time interval the corresponding surfaces; i.e. cone; of both objects must lie outside each other. It is possible to consider one object as a point and similar to classical approach taken in robot motion planning, enlarge another object with the same size. In this case, the problem becomes easier where the point should be outside of a polygon. With this trick linear conditions introduced by Schouwenaars et al. [33] can be used:

$$
\begin{aligned}
x_p^i - x_q^i &\geq d_x - R.c_{pq1}^i \quad and \\
x_q^i - x_p^i &\geq d_x - R.c_{pq2}^i \quad and \\
y_p^i - y_q^i &\geq d_y - R.c_{pq3}^i \quad and \\
y_q^i - y_p^i &\geq d_y - R.c_{pq4}^i \quad and \\
\sum_{k=1}^{4} c_{pqk}^i &\leq 3
\end{aligned}
$$

(2)

where d_l is the safety distance in direction l, c_{pqk}^i are a set of binary variables (0 or 1) and R is a positive number that is much larger than any position or velocity to be encountered in the problem.

4.2 Turn and Velocity Condition

It is possible to define other conditions to constrain the rate of turning (α_{max}) and changing velocity (Δ). Turn condition can be defined with the help of coordinates. Assuming space-time is small, linearization may apply. Other linear equations are suggested by Richards and How [30]. The velocity conditions can be derived easily as linear function from parameters.

$$
\boxed{\begin{array}{c}
\dfrac{x_p^{\,j} - x_p^{\,i}}{y_p^{\,j} - y_p^{\,i}} \le \alpha_{max} \\[2mm]
v - v^{*} \le \Delta
\end{array}}
\tag{3}
$$

4.3 Example

This example demonstrates that the suggested method forms an acceptable collision-free path for two boats. Figure 2 shows current locations of the boats, destinations, and space grids. The time axis is perpendicular to the space. Minimum distance optimality condition results to straight line paths to the destinations which clearly lead to a collision. In this example, the control parameters of the left vehicle are optimized.

Let minimum speed, maximum speed, fixed time interval, and maximum deviation angle (off-route angle) be 12 m/s, 30 m/s, 20 sec., and 5 degrees, respectively. It can be easily seen that the approximate envelope of cones with that deviation angle is a rectangle in 2-dimensional and a cylinder in 3-dimensional space. Figure 3 shows the result of optimization with equal weight for all parameters. As can be seen, only velocity of the left boat at collision time is reduced without any significant change in direction. In order to reach a minimum time trajectory or maximum traveling with a

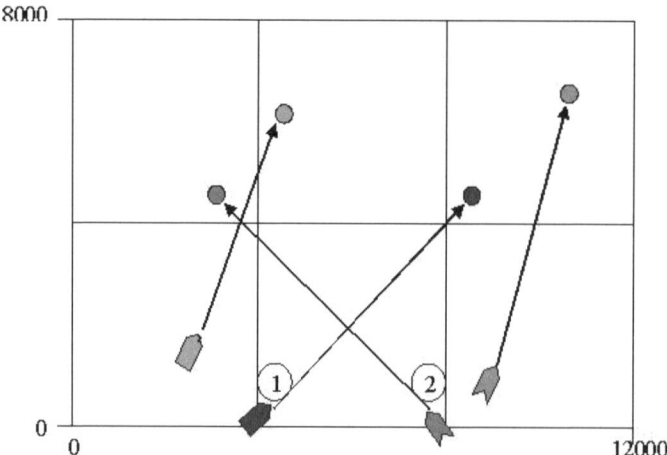

Fig. 4. The trajectories of two intersecting boats. Accident will occur at fifth time interval.

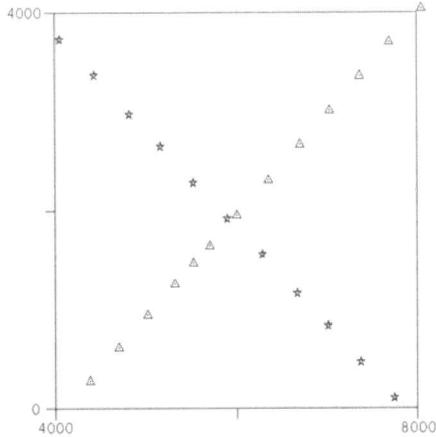

Fig. 5. The designed trajectory for the left boat when all parameters are considered with equal weights

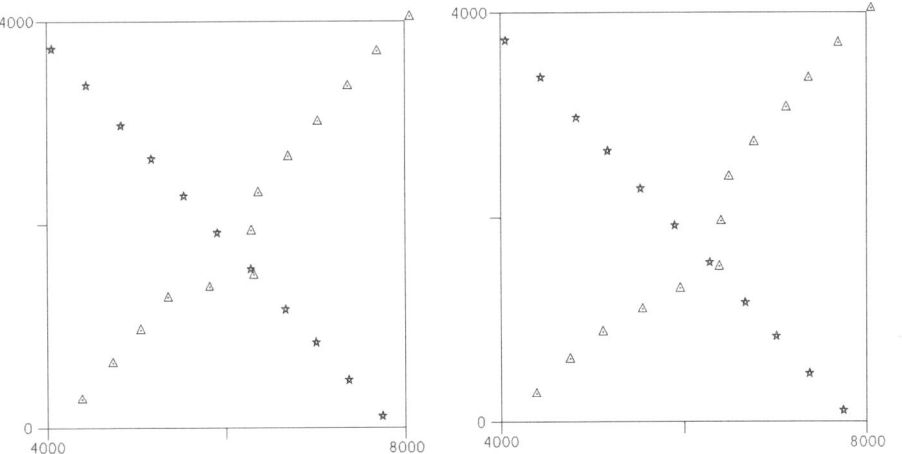

Fig. 6. The designed trajectories with different turn conditions and priority of the velocity

fixed amount of money in our scenario, high weights for velocity are defined and the results are shown in the Figure 4.

By this method it is not necessary to assume that target point and the altitude are fixed. In each space-time cell some new object can appear. Due to linear formulation, this approach may be used in real or fast-time systems.

5 Conclusion and Further Work

This paper addressed the collision-free path problem in the context of a limited resources mobile GIS environment. We have demonstrated that concerns to mobile

GIS theory can be addressed profitably in terms of the partition and conquer idea. It is based on partitioning space-time into small parts, solving the problem in those small cells and connecting the results with each other to find the final result. The reasons behind are clear. The problems can be solved easier and many things are predictable at a small part of space-time. Then, a logic-based framework for representing and reasoning about qualitative spatial relations over moving agents in space and time was derived. We provide convincing evidence of the usability of our suggested method by demonstrating how it can provide model for routing and navigation. A mathematical programming formulation has been proposed and simulated by an example to express optimal or near optimal collision-free path under the framework of such partitioning paradigm.

One important possible application of suggested methodology as our further work is mobile wayfinding services. It is based on the suggested method because wayfinding is an ordered presentation of the needed information to access an environment. It can be done in small parts as far as reaching to the desired point. A detailed uncertainty modeling for partitioning method and solving inverse problem, i.e., to determine the size and other characteristics of small parts based on the given information about needed precision, resource constraints, etc. are also among our future work.

Acknowledgements

The first author is partially supported by Ministry of Science, Research, and Technology of Iran under grant 805935. Thanks to Dr. Winter (University of Melbourne) for his comments.

References

1. Born, M.: "Natural Philosophy of Cause and Chance", Dover Publications, New York, 1949.
2. Caduff, D. : "Sketch-Based Queries In Mobile GIS Environments", Spatial Information Science and Engineering, Maine, University of Main: 114, 2002.
3. Cohn, A. G. and Hazarika, S. M.: "Qualitative Spatial Representation and Reasoning:an Overview", Fundamenta Informaticae, 43, pp. 2-32, 2001.
4. Chon, H., Agrawal, D. and Abbadi, A. E.:" Query Processing for Moving Objects with Space-Time Grid Storage Model", Dept. of Computer Science, University of California, No: 2001-15, 2001a.
5. Chon, H., Agrawal, D. and Abbadi, A. E.:"Storage and Retrieval of Moving Objects". Proceeding of *International Conference on Mobile Data Management.*, 2001b.
6. Chon, H., Agrawal, D. and Abbadi, A. E.:"Using Space-Time Grid for Efficient Management of Moving Objects". Proceeding of *MobiDE*, 2001c.
7. De Berg, M., Van Kreveld, M., Overmars, M. and Schwarzkopf, O.:"*Computational Geometry-Algorithms and Application"*,(Berlin, Springer-Verlag, 1997.
8. Erwig, M. and Schneider, M., 1999, The Honeycomb Model of Spatio-temporal partitions. In *Spatio-Temporal database management*, (Edinburgh, Scotland, Springer), PP. 39-59.

9. Ferscha, A. and Hoertner, H. and Kotsis, G.: "Advances in Pervasive Computing", Austrian Computer Society, 2004.
10. Forman, G. H. and Zahorjan, J.," The Challenges of Mobile Computing". *IEEE Computer*, **27** (4), 38-47, 1994.
11. GAMMA, http://www.cs.unc.edu/, Accessed: June 2006.
12. Garg, V. K. and Mittal, N., "Computation Slicing: Techniques and Theory". Proceeding of *DISC 2001*, Lisbon, Portugal, PP. 29-78, 2001.
13. GIS-LOUNGE, http://gislounge.com/ll/mobilegis.shtml, Accessed: June 2006.
14. Lewis, D.: "Causation", Journal of Philosophy, 70, pp. 556-567, 1973.Khatib, O., "Real-time obstacle avoidance for manipulators and mobile robots". *International Journal of Robotics Research*, **1** (5), 90-98, 1986.
15. Kaufmann, W. J.: "Relativity and Cosmology", 1966.
16. Kuchar, J. K. and Yang, L. C., "A Review of Conflict detection and Resolution Modeling Methods", *IEEE Transactions On Intelligent Transportation Systems* (December), 2000.
17. Latombe, J.-C., "*Robot Motion Planning*", Kluwer academic Publishers, 1991.
18. Laurini. R., "An Introduction to TeleGeoMonitoring: Problems and Potentialities", In Atkinson P. and Martin D.:"Innovations in GIS", Taylor & Francis, 2000
19. Li, L., Li, C. and Lin, Z., " Investigation On the Concept Model Of Mobile GIS". Proceeding of *Symposium on Geospatial theory, Processing and Applications*, Ottawa, 2002.
20. Malek, M. R., " Motion Modeling in GIS". Proceeding of *GEOMATIC 82*, Tehran, Iran, National Cartographic Center, 2003.
21. Malek, M. R.,"A Logic-based Framework for Qualitative Spatial Reasoning in Mobile GIS Environment", Lecture Note in Artificial Intelligence, Vol. 3066, pp. 418- 426, 2004.
22. Malek, M. R., Delavar, M. R. and Aliabady, S., "A Mobile Computing Approach for Rescue", proceeding of the 1st International Conference on Integrated Disaster Management, Tehran, January 2006, (in Persian).
23. Malek, M. R., Delavar, M. R. and Frank A.U., "A Logic-Based Foundation for Spatial Relationships in Mobile GIS Environment", Proceeding of 2nd Intenational symposium on LBS & Telecartography, Austria, Vienna, 2006.
24. McClosky, M.: "Naive theories of motion" In: Gentner D. and Stevens S. (Editors): "Mental Models", Hillsdale, New Jersey, Lawrence Erlbaum, 1983.
25. Mittal, N., "Techniques for Analysing Distributed Computations", Department of Computer Science, Austin, USA, The university of Texas, 2002.
26. Nivala, A. M. and Sarjakoski, L. T.: "Need for Context-Aware Topographic Maps in Mobile Devices", In: Virrantaus, K. and Tveite, H.: *ScanGIS'2003*, Espoo, Finland, 2003.
27. Pallottino, L., Feron, E. and Bichini, A., "Mixed Integer Programming for Aircraft Conflict Resolution". Proceeding of *Guidance, Navigation and Control Conference*, 2001.
28. Pallottino, L., Feron, E. and Bichini, A., "Conflict Resolution Problems for Air Traffic Management systems Solved with Mixed integer Programming". *IEEE Transactions On Intelligent Transportation Systems*, **3(1)** (March), 3-11, 2002.
29. Richards, A., How, J., Schouwenaars, T. and Feron, E., "Plume Avoidance Maneuver Planning Using Mixed Integer Linear Programming". Proceeding of *AIAA 2001*, 2001.
30. Richards, A. and How, J. P.,"Aircraft Trajectory Planning with Collision avoidance Using mixed Integer Linear Programming". Proceeding of *American Control Conference 2002*, 2002.
31. Sang, J.: "Theory and Development of GPS Integrity Monitoring System", PhD Thesis, Queensland University of Technology, 1996.
32. Satyanarayanan, M., "Fundamental Challenges in Mobile Computing". Proceeding of *ACM Symposium on Principles of Distributed Computing*, 1995.

33. Schouwenaars, T., Moor, B. D., Feron, E. and How, J., "Mixed Integer Programming For Multi-Vehicle Path Planning". Proceeding of *European Control Conference 2001*, 2001.

34. Verma, T. S.: "Causal Networks: Semantics and expressiveness", In: Sacher R. and Levitt T.S. and Kanal L.N. (ed.s):"Uncertainty in Artificial Intelligence", Elsevier Science, 4, 1990.

35. Van den Bergen, G., *"Collision Detection in Interactive 3D Computer Animation"*, Eindhoven, Eindhoven University of Technology, 1999.

36. Wolf, H.,"The Helmert block method, its origin and development", Proceeding of *Second International Symposium on Problems Related to the Redefinition of North American Geodetic Networks*, Arlington, PP. 319-326, 1978.

37. Wolfson, O., Jiang, L., et al., "Databases for Tracking Mobile Units in Real Time", Proceeding of *Database Theory- ICDT'99, 7th International Conference, LNCS 1540*, PP. 169-186, 1999.

38. Zaslavsky, A. and Tari, Z., " Mobile Computing: Overview and Current Status", *Australian Computer Journal*, **30**, 1998.

39. Zhao, Y., *"Vehicle Location and Navigation Systems"*, Boston, Artech House, 1997.

Improving Archaeological Heritage Information Access Through a Personalised GIS Interface

E. Mac Aoidh, A. Koinis, and M. Bertolotto

School of Computer Science and Informatics, UCD Dublin, Ireland
{eoin.macaoidh, aggelis.koinis, michela.bertolotto}@ucd.ie

Abstract. Current archaeological heritage dissemination systems do not take full advantage of available modern technology. For example, the linking of archaeological findings to their geographical surroundings is a functionality offered by few systems. Given the diversity of webusers, a personalised presentation of the information would be desirable. The TArcHNA GIS architecture offers dynamically tailored spatial and non-spatial information to its users. The vast quantity of archaeological heritage information in the system is filtered to suit each individual, based on user models created by previous interactions with the system. The heritage information is made accessible via a personalised map interface. User interactions are captured implicitly, without the users knowledge. The system is designed to operate on both mobile and desktop devices enhancing the accessibility, and the user's appreciation of archaeological heritage.

1 Introduction

Public access to information on archaeological heritage is somewhat restricted. In many cases the dissemination process of such information through current systems is overly simplistic and perhaps lacks the benefits of modern interfaces. Through the use of modern technology combined with an extensive knowledge of the Etruscan history and archaeological findings, the TArcHNA (Towards Archaeological Heritage New Accessibility)project aims to develop new models and tools for accessing archaeological heritage. 'Tarchna' is the Etruscan name for the city of Tarquinia.

TArchNA advances current systems use of static, non-personalised interfaces by integrating GIS (Geographical Information Systems) functionality into a system for disseminating cultural heritage information, which is capable of personalising both the information content and the display. The project goal is to enhance the user's appreciation of archaeological sites in Tarquinia during his visit to such a site, using the mobile TArcHNA GIS application, or in a museum or home setting by using the web-based version of the system. The provision of adaptable, easy-to-use, web-based and mobile systems, offering maps as an information portal to TArcHNA's archaeological information is our primary objective.

The archaeological information details burial tombs and findings within the tombs. The quantity of data in the system is continually growing as more tombs are discovered and documented. The spatial information contained in the

J.D. Carswell and T. Tezuka (Eds.): W2GIS 2006, LNCS 4295, pp. 135–145, 2006.

system's interactive map is the user's gateway to the growing repository of archaeological information. Clicking on locations and tombs on the map displays associated information from the database. TArcHNA uses off-the-shelf technology and industry standard formats to minimise the complexity of the system and its data. By dynamically personalising each user's interaction experience with the system, our contribution exploits GIS and personalisation, enhancing user's ease of access and increasing user's appreciation of archaeology and cultural heritage in a web-based or mobile environment. User interaction with the interface is continually monitored. This information is analysed to create appropriate user models which are then used to deliver personalised content to individual users.

The remainder of the paper is organised as follows, section 2 gives an overview of projects that address certain aspects bearing similarities to the TArcHNA project. This section also addresses the issues of user modelling and interface personalisation. Section 3 details the architecture of our system. Section 4 addresses user profiling and personalisation. Section 5 details future work.

2   Related Work

In this section we review related work pertaining to both similar heritage information systems, and selected systems which profile users to create user models.

2.1   Cultural Heritage Systems

The "Valley of the Shadow" [1] system archive contains original letters, newspapers, speeches, census extracts and other documents from two American communities between 1859 and 1870. Users can access a large variety of archived information from the era through the project web page, information that was previously only available to historians. The information was collected and is filed away according to type and subject matter. Users must sift through all this information with no help from the system in terms of information filtering or display. In addition, information is not cross-referenced i.e. photographs relating to a specific topic are not linked to newspaper articles or census documents of the same topic.

The "Theban Mapping Project" [2] aims to create a comprehensive archaeological database of Thebes, an important archaeological site with thousands of tombs and temples. The system stores information on these sites and makes them accessible to the public through 3D imagining, sketches and photos. These elements accompany a narration to present a complete description of each site. A detailed map of the region with each archaeological site depicted is provided. The system associates location and relevant information very effectively. Animated zoom and pan sequences between tombs leave the user in no doubt as to his overall location within the archaeological site. TArcHNA gives details of the tomb architecture, but also gives details of the findings within the tombs. The Theban system is confined to the architecture of the tombs. The information content provided by TArcHNA is much more extensive and detailed, thus we rely on personalization and collaboration amongst users to reduce complexity and present information logically to our users.

The ARCHEOGUIDE [3] system provides a tour of the archaeological site at Olympia in Greece using augmented reality, 3D-visualization, mobile computing, and multi-modal interaction techniques. It places emphasis in virtual reconstruction of the archaeological remains. The system consists of an on-site information server and a set of mobile units that are carried by visitors. A wireless local network allows the mobile units to communicate with the site information server. The system is aimed at users on-location. This limits the target audience to those visiting Olympia. The user can view a map of the site in order to have an idea about his location and the general area around him. However this map is merely a representational diagram of his location on the site. ARCHEOGUIDE is an effective tour guide system, however, its use of a 3D environment adds to the complexity of the project and highlights the use of specialised equipment. TArcHNA avoids the use of any specialised equipment, and promotes the dissemination of its information to all users by supporting both mobile on-location and remote access to its data.

Collectively, these systems reveal that an effective interpretation of GIS technology has not been achieved for the purpose of archaeological heritage information dissemination. The maps provided by these systems are very limited. The Valley of the Shadow project provides basic raster maps with no interactivity. The Theban system provides a map that can be used to retrieve information, however this is simply an animated picture to give the impression of interactivity. ARCHEOGUIDE's maps are also non-interactive, although the user's location is dynamically represented on the map as he moves through the site.

TArcHNA aims to disseminate its heritage information through the map interface, using the map as the focal point which connects all the information. The interactivity and usability of our map interface is the key to the provision of all archaeological information, as it is primarily accessed through the map interface. Maps are kept as legible and focused as possible by personalising the maps and the non-spatial information to suit a particular user's interests. User models are implicitly produced by the system for the purpose of personalisation.

2.2   User Modelling and Personalisation

Much work has been documented in the literature regarding user modelling. The motivation for this work is the potential improvement yielded in the collaborative nature of HCI [4]. The advancement of user modelling leads to an improved experience for the user. For a given system, the employment of user models allows the system to tailor the information returned to each individual user according to his user model. This has a very broad appeal across all applications from generating lists of suitable books for a user on Amazon, to delivering advertisements to a user based on the contents of the web pages viewed in his browser, by Google's advertising programmes. These services demonstrate the significant commercial potential for user profiling.

The development of adaptive systems, which dynamically adapt to the user, make the user's goal easier to achieve. User context must be taken into account to facilitate system adaption. Nivala and Sarjakoski [5] detail the contexts which should be considered when designing an adaptive map interface. At the heart of an adaptive system is a component which creates user models. Of primary interest to TArcHNA are methods which implicitly collect information about the user to generate such models.

CHEESE [6] tracks and records mouse movements on a web page. The authors argue that the user's positioning of the mouse on a web page subconsciously reveals his level of interest in a particular area of the page. This recorded information allows the system to create a user model which makes informed assumptions about the user's interests. Claypool et al's Curious Browser [7] records the user's actions to generate implicit ratings. Curious Browser focuses on mouse clicks and key press actions as the user browses the web. CHEESE and Curious Browser infer users' interests from non-spatial data.

CoMPASS [8] provides personalised maps to its users. It employs implicit profiling, monitoring a user's interactions with its spatial content. The map elements within the user's view frame are monitored; the level of zoom, number of features and interval between user actions are all monitored to implicitly determine interest. The information represented in CoMPASS is entirely spatial. Personalisation of the spatial data returned to the user is performed. The interface itself remains unchanged, in contrast to our GIS system, which presents both spatial map data and non-spatial information to the user. It is an ideal environment within which to compare and contrast the system's prediction accuracy of user's intentions with textual, non-spatial and spatial data. Our system is designed to explore users' mouse movements over spatial data and associated non-spatial data such as text and images in order to generate a personalised interface and dataset for a given user.

3   TArcHNA System

TArcHNA intends to develop new models and tools for accessing archaeological heritage. We have developed two applications incorporating GIS technology: a Web-based application, and a mobile PDA application. These applications share the same basic architecture and can each be used as an extension of each other, or as independent applications. The dataset is comprised of both spatial, and non-spatial data such as text, pictures, and movies.

3.1   Web-Based Architecture

The TArcHNA GIS GUI and functionality is based on OpenMap [9]; an open-source Java based mapping toolkit provided by BBN Technologies. It is programmed in Java and is designed to be embedded in a web page so that it is available to everyone who wishes to access it. The maps used by the application consist of spatial data and orthophotos of the selected area. The spatial data is organised in ESRI shapefile format and the photos in JPG format. All data files are stored in the directory within which the application is running in order to reduce access time. Map Properties Files (MPF files) are used to organise the data that describe the layers comprising a map and specify how the map is to be viewed. These data files are not the only primitive files TArcHNA GIS uses. Maps constitute part of the application's data, and are organised in a hierarchical fashion. Each child shows an area of its parent map in greater detail. The user obtains further detail as he moves further down the hierarchy. The user can move through this hierarchy, by clicking on active points in the map which provide the mechanism for jumping to the next map in the hierarchy. These locations are described in CSV files and are organised in layers.

Detailed narrations have been compiled by expert archaeologists at various levels of detail ranging from tomb overviews, to details on individual chambers of the tomb, to wall paintings and artefacts found within a given chamber. These narrations, accompanying pictures, movies and sound files are all available to the user through the system's interface by interacting with the map. When a spatial object is selected on the map, we can conclude that the user requires further information about that object. In this case the application would send a query to the database. The query, sent using SOAP calls, would return a result from the database in XML which would be parsed and displayed to the user in the information browser. However, due to the large volume of information available to the user, we perform a transparent personalisation step at this point to customise the users display. This is described in detail in section 4.

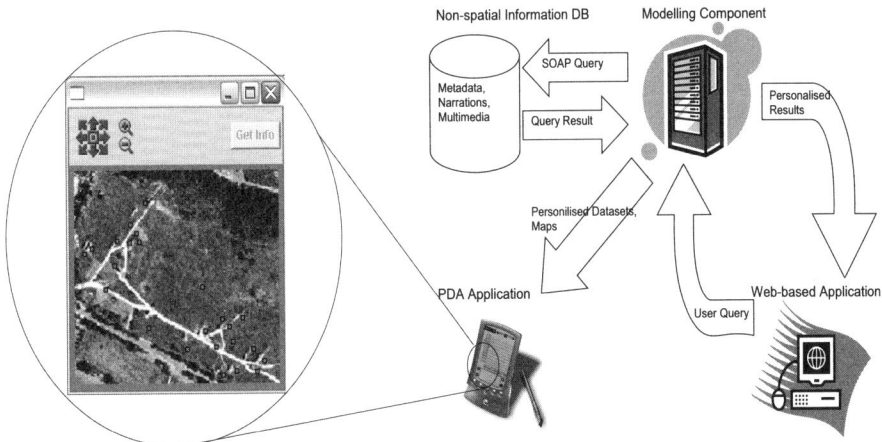

Fig. 1. System Architecture with enlarged PDA screenshot

The web-based application takes advantage of the powerful processors on desktop computers, allowing for a wide arrange of functions and user interaction. The basic functionality offered by the web-based application allows the user to:

- Navigate around the maps by using zoom and pan buttons.
  Access all available maps by choosing a map from a list or by clicking relative locations on the map.
- Access information on objects shown on the map by clicking them.

- Make area-select queries. The user can specify an area on the map, and the application retrieves information on the objects in that area.
- Create a list of archeological items that are of interest to the user. This list can be created and exported to the mobile application.

3.2 Mobile Application

The mobile application is a modified version of OpenMap designed to work within the compact environment of the PDA. It is programmed in the Java2ME Personal Profile environment as it provides a stable, versatile environment to build mobile applications. The mobile application is designed for use while in the field, visiting archaeological sites. It offers the user the same basic functionality as the web-based application, with some simplifications and adjustments to suit the smaller screen, lesser CPU power and use of a stylus. In addition the users exact position and overall route can be represented on the map using a GPS receiver. The mobile application uses the same data as the web-based application. The shapefiles and orthophotos are stored on a storage card in the device. The MPF is used again to organise the map layers, however, the use of the device in a specific area eliminates the need to accommodate a map hierarchy. The user is simply given a list of detailed maps in his immediate area from which he can choose as he roams. In order to keep implementation as simple and reliable as possible, this application does not communicate wirelessly with the information database. A pre-made dataset from the database is loaded onto the device before the user leaves base. The contents of the pre-made dataset are determined by the user's profile as described in section 4.

4 Personalisation

The system's data repository is continually growing as new tombs are discovered and documented. In order to make the data more manageable to the user it must be filtered, reduced and simplified. Many proprietary systems currently offering 'personalisation' merely allow the user to change the colour of the interface components, or fonts. Our systems concept of personalisation is much deeper than this. The system architecture is the foundation for an intelligent system dealing with both spatial and non-spatial data. It monitors each user's interactions with the system and builds an implicit user model as they interact with the system. The manner in which this information is captured, and will subsequently be utilised to create a personalised user experience is central to our unique approach.

4.1 Intended Users

There are three broad classifications of intended users of the TArcHNA system: 1) Novice tourist users 2) returning advanced users and 3) expert users. Tourist users are treated as a group of one-time users, however it is expected that as a group they will have similar interests and preferences. To this effect, we treat

Fig. 2. Web-based application screenshot

them as a collaborative family of users, where each user contributes to create an overall, generalised, unified model of a tourist. Their use of the system is expected to be once off, either on a mobile system on-location, or a desktop system at a kiosk in a museum. Returning users on the other hand are likely to have more interest in the region than an average tourist. Such users include archaeologists and historians. Their use of the system occurs both in the field as a mobile GIS, and on remote desktops as a web-based GIS. Each of these users are modelled individually for their own unique profile. They are also more broadly classified into advanced and expert collaborative user families. The expert users of the system include the archaeologists who update the information in the system.

4.2   User Model Construction

By storing user preferences in user models, the system can return data and functionality with increased relevance for a particular user. In order to learn the user's preferences, it must continually monitor his actions and interpret them to infer likes and dislikes. Interaction with the system falls into two distinct categories; spatial and non-spatial interactions. At present our system focuses on spatial interactions.

Mouse actions are continually logged in a database. The time a mouse movement or click takes place, the duration of each action, and its location in relation to the spatial information displayed (tomb location) are all recorded. Further to this, zoom and pan actions, explicit searches and object placement within the frame of the map are all taken into account. Mueller et. al [6] have shown user's mouse actions to be indicative of their interests with non-spatial data. Early indications of our implementation are that the user's subconscious mouse actions with spatial data also positively disclose their implicit interests.

TArcHNA GIS, at an early stage of development, focuses on the collection of spatial data, however it is intended that this data will be combined with the information gleaned from interactions with non-spatial data to strengthen any assumptions made. Data such as hyperlinks followed in the documents returned to users, keywords indicating their subject matter, and the duration of viewing a page are all expected to be taken into account. While the user monitoring structure is under development, the interpretation process of the user's actions and subsequent personalisation by the system is still at a conceptual stage.

4.3   Personalisation Functionality

As described in section 3.1, the dataset for the system is stored remotely. Requests for information are acknowledged by the interface, formatted as XML and are sent to the database. They are sent via the modelling component which acts as an intermediary between the client's interface and the vast quantity of information stored on the remote server. It dynamically builds personalised requests for information, and sends them to the remote server as XML on the client's behalf. The remote server fetches the required data, and formats a reply in XML. This XML file is parsed initially by the user modelling component, which screens the remote server's reply to ensure it suits the needs of the user in question. The implementation of this section remains at an early stage while the data collection functionality is being developed. However we expect it to develop using well-established techniques ([10,11]) for non-spatial web data, following the basic steps of data preprocessing, pattern recognition (e.g. K-Nearest-Neighbors and association rule mining), and pattern analysis. These techniques, subjected to appropriate modification for spatial data form the basis for our system's data personalisation component.

Further personalisation of the information may need to be carried out at this stage such as the elision of extraneous data, [12] by placing it on (a) subsequently linked page(s). The personalised response is then passed to the client side, which parses the XML, and visually renders it for the user as HTML, displaying links to other relevant information, pictures, videos and sound files. The user then interacts with this newly reduced data and/or further spatial data. Throughout the duration of the interaction his actions are continually logged as described in section 4.2, and stored on the modelling server. We have conducted a skeleton implementation of this personalisation functionality, however, it remains largely conceptual and subject to change as the user modeling component develops.

4.4   Personalisation of Display Content

HCI in GIS is complicated by poorly designed interfaces, and information overload. Our system architecture is designed to both personalise the interface; displaying the map data appropriately, giving the user access to the tools he needs, and personalise the dataset, reducing information overload. The user models created by the system provide information about a user's preferences for data display and content.

Interface Personalisation. Screen real estate is an expensive commodity. An interface must be arranged efficiently to avoid clutter, but must also display enough data and functionality to reduce the number of steps a user must follow to complete a given task. Different user groups have distinctly different information display and manipulation needs. For example, our tourist family of users are mostly passive, absorbers of the information supplied to them by the system. They have little need to manipulate the data, in contrast to an advanced user, who might need any number of manipulation tools to work with the data returned to him.

We aim to provide a single adaptive interface to all users which will provide them with the functionality they most likely require, and the relevant data displayed in the most desirable format to suit their needs. The user model defines the personalised format of the interface. Functionality that the user made use of, or avoided in the past is weighted accordingly to determine the functions available in the quick access toolbar on the interface. Moreover, each user may favor various components of the interface to varying degrees. E.g. A user with a narrow spatial aptitude may find it difficult to interpret the spatial data presented in the maps, and might favor a smaller map window, with more emphasis being placed on the text window, or a larger display window for pictures.

Dataset Personalisation. Information overload is a common problem associated with the wealth of information available to computer users. A web search for the term "car" returns over 1.5 million results. This is simply too much information for one person to handle. Web search engines deal with this problem by ranking the results. In theory the first result should be the most authoritative site for the term "car", with the last site being the least authoritative. TArcHNA has a large quantity of spatial and non-spatial data available to it. Returning all of this information to any user would cause information overload.

Information overload is tackled in TArcHNA by reducing the dataset returned to the user, making it easier for the user to find what he is looking for. Non-spatial data is handled similarly to the web search analogy; it is ranked based on its suitability for a given user. For example, if a users profile indicates that he is interested in ceramic artefacts, when he searches for tombs from a certain era, those with information on ceramics will be given more prominence on the display.

Spatial data can also be ranked by suitability for the user. Similarly to non-spatial personalisation, it is carried out based on the information in the user profile. For example, if this information indicates that the user is particularly interested in tombs of type x in a given locality, then tombs of type x in the immediate area will have a greater relevance ratio than other tombs, and will be displayed appropriately.

User Families. Generating each user's interface and dataset dynamically from their user profile presents a problem when a new user wishes to use the system.

Which tools should be included on the interface? How should the display components be laid out? What data should be given a higher preference? - The classic cold start problem [13]. In order to solve this, users are further grouped into collaborative "families" of users as mentioned in section 4.1. The tourist family model is a representation of the typical settings that suit the average tourist. The more tourists use the system, the more accurate this average becomes.

Individual models are not stored for tourists, as they are considered one-time users of the system. Advanced and expert users have individual models which are maintained by the system, these models are grouped together to form collaborative models to suit the average advanced or expert user respectively. When a new user wishes to interact with the system, the interface suitable for his particular user family is initially presented to him.

5   Future Work

The architecture for our system has been put in place as described and is subject to constant modification as it develops. While the system is fully functional, the personalisation component is yet to be completed. This component awaits further development to fully avail of all the information currently collected from each user's interactions in order to personalise the display and content accordingly.

As the system's development is still on-going, there is an extensive amount of future work we wish to cover, and a number of caveats mentioned in the literature of which we must take heed. The narrowing of a user's options is an example of such a caveat. Both interface and dataset personalisation inevitably leads to the narrowing of user's options. If a user continually uses function X and Y from the quick access toolbar, function Z might be removed due to disuse, refining his choice of functionality. If the user's dataset is also continually refined, after a while his overall scope will be greatly diminished. As a temporary solution to this problem the user has access to a full list of functionality and spatial data layers from which he can add any additional functionality or data to his interface. We expect to resolve this problem with a more intelligent algorithm and a more concrete implementation.

Following the completion of the personalisation functionality, we aim to make a full assessment of the system in order to document its strengths and weaknesses. Our work will contribute to the GIS community by strengthening spatial data recommendations made to the user based on inferred user interests from spatial data interactions.

Acknowledgement. The support of the TArcHNA project, funded under the EU Culture 2000 Programme is gratefully acknowledged.

References

1. Valley of the shadow. http://valley.vcdh.virginia.edu/.
2. Theban mapping project. http://www.thebanmappingproject.com/.
3. Archeoguide. http://archeoguide.intranet.gr/.

4. G. Fischer. User Modeling in Human-Computer Interaction. In *Proceedings of the 10th Anniversary issue of User Modeling and User-Adapted Interaction*, 2000.
5. A Nivala and L.T. Sarjakoski. Need for Context-Aware Topographic Maps in Mobile Devices. In *Proceedings of the 9th Scandinavian Research Conference on Geographic Information Science ScanGIS2003*, pages 15–29, Espoo, Finland, June 4-6 2003.
6. F. Mueller and A. Lockerd. Cheese: Tracking Mouse Movement Activity on Websites a Tool for User Modeling. In *Proceedings of the Conference on Human Factors in Computing System (CHI'2002)*, 2002.
7. M. Claypool, P. Le, M. Waseda, and D. Brown. Implicit Interest Indicators. In *Proceedings of the International Conference on Intelligent User Interfaces (IUI'01)*. ACM, January 14-17 2001.
8. D. Wilson, J. Doyle, J. Weakliam, M. Bertolotto, and D. Lynch. Personalized Maps in Multimodal GIS. *International Journal of Web Emerging Technology*, In Press, 2006.
9. Openmap. http://openmap.bbn.com/.
10. B. Mobasher and H. Dai. A Road Map to More Effective Web Personalization: Integrating Domain Knowledge with Web Usage Mining. In *Proceedings of the International Conference on Internet Computing 2003 (IC03)*, Las Vegas, USA, 2003.
11. B. Mobasher, H. Dai, T. Luo, and M. Nakagawa. Effective Personalization Based on Association Rule Discovery From Web Usage Data. In *Web Information and Data Management*, 2001.
12. T. W. Bickmore and B. N. Schilit. Digestor: Device-independent Access to the World Wide Web. In *Selected papers from the sixth international conference on World Wide Web*, pages 1075–1082, Santa Clara, California, USA, 1997.
13. D. Maltz and K. Ehrlich. Pointing the way: Active Collaborative Filtering. In *Proceedings of the SIGCHI conference on Human factors in computing systems*, pages 202–209, Denver, Colorado, USA, 1995.

Designing Adaptive Spatio-temporal Information Systems for Natural Hazard Risks with ASTIS

Bogdan Moisuc, Jérôme Gensel, Paule-Annick Davoine, and Hervé Martin

LSR-IMAG Laboratory, 681, rue de la Passerelle, Domaine Universitaire,
38 402 Saint-Martin d'Hères, France
{bogdan.moisuc, jerome.gensel, paule-annick.davoine,
herve.martin}@imag.fr

Abstract. This paper presents ASTIS, a framework for the design and generation of adaptive spatio temporal information systems for the historical study of
natural hazard risks. ASTIS is based on a modular architecture in which every
module can be personalized by the designer in order to meet the needs of different kinds of users. Personalizations are performed through a model-driven
approach, each module is generated from specific models, conceived by the designer. A data management module allows personalizing the content of the
application via data viewpoint mechanisms. A presentation module allows designing personalized interactive visualizations. Finally, an adaptation module
is in charge of performing the appropriate personalizations at runtime, in order
to adapt both the content and the presentation of the information to the user
characteristics.

1 Introduction

Spatio-temporal information systems are frequently used nowadays for natural hazards
management. Designers of spatio-temporal information systems dedicated to natural
risks (called hereinafter ISNR) have to face a series of difficult problems. Managing
spatial and historical data about natural hazards means tackling issues related to the representation and the visualization of complex and three-dimensional (spatial, temporal
and thematic) information. Such issues typically require capabilities of multidimensional visualization of complex information, completed with interactive mechanisms,
which allow the execution of visual queries on each dimension of the information [5].
Moreover, although natural risk management is a multidisciplinary field and requires
applications to meet the needs of various kinds of users (earth scientists, regional planners, decision makers, etc.) with different computer literacy levels, existing systems are
often targeted only towards the expert users. In order to improve their usability, ISNR
should be adaptable to the numerous and relevant characteristics of the users, like their
interests, roles, objectives, levels of expertise, etc.

However, very few researches are oriented towards the design of adaptive spatial
or spatio-temporal systems. Most known approaches come from the domain of mobile
tourist guides. Most approaches (e.g. [1], [10]) provide adaptation mechanisms for the
location of the users. Some of the works focus on adaptation mechanisms to the limited capabilities (in terms of display and memory size) of the mobile devices [14].

J.D. Carswell and T. Tezuka (Eds.): W2GIS 2006, LNCS 4295, pp. 146 – 157, 2006.

Few researches take into account the user profile for the adaptation. Nivala and Sarjakoski [10] adapt the style of the maps to the age and skills of the users. Zipf *et al.* [18] proposed an adaptive framework that takes into account the cultural background of the users in order to adapt the graphic styles of the displayed maps. Unfortunately, most of these researches handle only some aspects of presentation adaptation, while content adaptation to the users is ignored. To our knowledge, only Nivala and Sarjakoski [10] provide some content adaptation mechanisms, under the form of *use cases* (adapting the map data to the task performed by the user) that users can choose, which are hardcoded within the application structure. Because adaptability is not taken into account from the design phase, most existing approaches provide only superficial personalization mechanisms. Moreover, the adaptation mechanisms are intermingled with the application logics, which makes their reuse or extension very difficult [13].

We focus on the design of adaptive STIS applications, i.e. applications that adapt themselves to the users without requiring their intervention ([4]). We argue that adaptation and adaptivity should be taken into account from the design phase, and that adaptation mechanisms should target both the content and the presentation of the STIS applications. This paper presents ASTIS (*Adaptable Spatio-Temporal Information System*), a modular framework that allows designers to conceive and generate adaptive and interactive ISNR. This framework relies on a previously introduced architecture, called GenGHIS [8], which allows generating ISNR designed for a specific application domain by creating a specific data model. GenGHIS relies on the use of a object-based knowledge representation system extended with space and time called AROM-ST [11]. The interface of GenGHIS allows users to visualize the data in an interface displaying three frames, corresponding to the spatial, temporal and thematic views of the data. ASTIS extends the GenGHIS architecture with mechanisms that enable the personalization of both data and presentation. The data management module has been extended with data viewpoint mechanisms that enable content personalization by deriving personalized data schemas from the initial data schema of the application. A presentation module allows designers to create personalized presentations that integrate well-proven dynamic and interactive visualization techniques like *brushing* and *linking* ([2], [3], [9], [6]). ASTIS also includes an adaptation management module, which automatically performs at runtime the personalizations required by the designer. We follow a model driven approach for personalization, each module is generated from specific models (for data, presentation and adaptation) created by the designer. In order to assist the ISNR designer in the modeling task, ASTIS offers, for each module, generic models that they can extend by specialization and/or by instantiation. This way, the ISNR designers can generate interactive STIS applications that can adapt themselves to the end-users by relying only on a modeling approach.

This paper is organized as follows: in section 2, we present the general structure of the application and its principles. Section 3 describes the general aspects of adaptation management. Section 4 and 5 detail our approach for generating adapted contents and adapted presentations. Section 6 concludes and gives some directions for future development.

2 Overview of the ASTIS Framework

Our approach relies on creating open modules, able to perform generic tasks of data acquisition, management, analysis, interrogation and visualization of spatial, temporal and thematic data. In order to adapt each module to the particular needs of their application, ISNR designer create models that describe the structure and the dynamic behavior of the application. In order to assist the designers, for each module we propose general models, from which they can design their specific models by specialization or instantiation.

ASTIS is composed of three main modules (see Fig. 1):

1. The data module has two main functions: *i*) it is in charge of the storage, interrogation and analysis of data in conformity with data models created by the designers and *ii*) it allows designers to instantiate their data models by data acquisition processes from external sources (different standard formats are supported, like mif/mid, dbf, xls, shp, etc.), in order to populate the knowledge base that serves as support for the application.

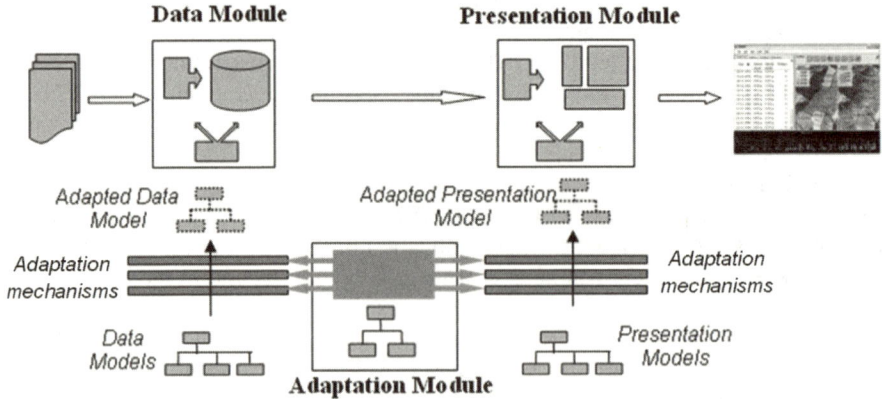

Fig. 1. General architecture of the ASTIS framework

2. The presentation module has two main functions, performed in conformity with the presentation model specified by the designers: *i*) it allows to transform data from the storage format into the presentation format applying the required style options and, possibly, analysis options and *ii*) it displays the presentation data through visualization interfaces structured according to the presentation model, containing synchronized spatial, temporal and thematic frames, and their respective widgets (buttons and controls), allowing end users to execute visual queries.

3. The adaptation module controls the two other modules, adapting the application to the profiles of the end-users in terms of content and presentation. It has two main functions: *i*) it allows designers to describe the users of the application (groups or individuals) and the way in which the application should adapt to them (through adapted data and presentation models), by instantiating the generic adaptation

model, *ii*) at runtime, it performs automatically (without any action from the end-users) the adaptations described by the designer in the instantiated adaptation model, by choosing the appropriate data and presentation models and/or by applying them the appropriate transformations.

In the next sections, we give an overview of the adaptation process and then we detail our approach for content and for presentation adaptation.

3 Adaptation Process

Adaptivity aims at providing end users with the appropriate informational content, under the appropriate form, with regard to their objectives, profession, interests, experience, cultural background, etc. All these user characteristics result in different informational needs and should be used in order to determine which relevant data are to be presented to them. We group all these characteristics under the concept of **user viewpoint**. The viewpoint defines a particular perspective on an application of a certain group of users, as a result of their objectives, profession, etc. Other targets for the adaptation should be considered as well. The **expertise level** defines the experience of users with the application, with the field of the application and with computer tools in general. This should be reflected by the complexity of the visualizations and of the functionalities that are offered to different users.

In order to adapt to each individual user, the system must store some information about them, which is described using an object oriented user model. User descriptions are considered at different aggregation (and abstraction) levels. The concept of group (class *Group* in Fig. 2) simplifies the work of the designers and gathers the common characteristics for sets of users. One group can be composed of other groups, thus it is possible to specify more or less general methods of adaptation. For example, the designer can define a group of "geologists" (this implies a certain viewpoint on the data and certain visualization methods), composed of a group of advanced users (in terms of skill computer literacy) and a group of beginners, in order to differentiate the complexity of the functionalities available to the two sub-groups.

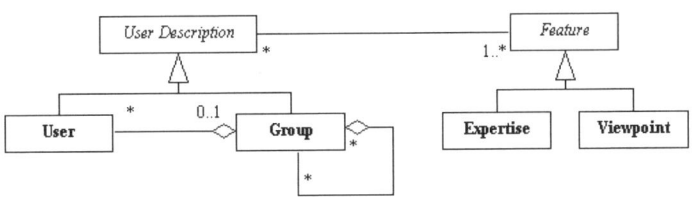

Fig. 2. The user model of ASTIS

In ASTIS, adaptivity is based on a three-stage process. First, the designer defines the initial data model and the initial presentation model of an application. Then, he has to define the adaptation model. This means *i*) defining "views" of an application, by describing a set of informational contents (viewpoints on the data) and a set of presentations and *ii*) defining the user profiles, by defining the potential groups of users and by associating to each group their respective viewpoints corresponding to their preoccupations. *iii*) at runtime, the system filters the set of possible "views", selecting for each user the "views" that correspond with their needs. We detail herein-after the description of viewpoints (content adaptation) on data and, respectively, on visualizations (presentation adaptation) within ASTIS.

4   Content Adaptation Through Data Viewpoints

Content adaptation is centered around the concept of **data viewpoint,** which defines a particular perspective on the application of a certain group of users, as a result of their objectives, profession, etc. In ASTIS, defining a data viewpoint means deriving from a unique and global data schema of an application, multiple and partial schemata, containing information concerning different perspectives (see Fig. 3) Thus, a data viewpoint defines: *i*) a data schema containing "new" classes, associations, attributes and roles (see, respectively, classes *Viewpoint Class*, *Viewpoint Association*, *Viewpoint Attribute* and *Viewpoint Role* in Fig.3) and *ii*) a mapping for entities from the base data schema to the derived one (see the relations *base class* and *base association* in Fig.3).

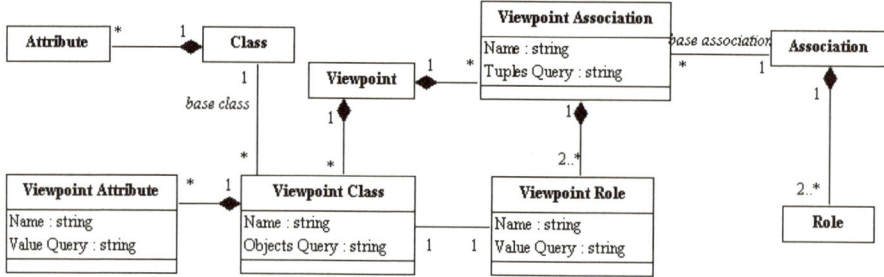

Fig. 3. Viewpoint structure in ASTIS

The designer can use several mechanisms for deriving viewpoints:

1. Masking classes and associations, attributes and roles;
2. Masking irrelevant objects and tuples, by using filtering queries that allow only the relevant ones to be included (the *Objects Query* attribute for the class *Viewpoint Class* and, respectively, the *Tuples Query* attribute for the class *Viewpoint Association*);
3. Renaming classes, associations, attributes and roles;
4. Merging classes that are related through associations into only one class (the *Value Query* attribute in the *Viewpoint Attribute* class allows retrieving the value of the attribute from attributes of classes linked to the base class), or splitting one class into several ones (by defining several classes from the same base class);
5. Defining new classes, using queries to cluster objects of the same type (by using the *Object Query* attribute in class *Viewpoint Class*).

Let us illustrate on a short example how the above mechanisms can be used. Let us suppose that a designer is creating an information system for landslide risks. Starting from the base data schema (see Fig. 4a) the designer wants to describe a "regional planner" viewpoint (see Fig. 4b). This requires masking irrelevant classes and associations, for instance, geological information about landslides (class *Geological Layer* and the corresponding association) will simply be left out.

Further on, let us suppose that in the base data schema, landslides are recorded not with their exact location instead they are included in larger probable landslide location sites (see class *PLL Site* in Fig. 4a). However, this aspect should be made transparent for regional planners. As a simplification, the "regional planner" viewpoint merges the two classes (*Landslide* and *PLL Site*), it leaves aside (masks) the *Serial* attribute and defines the location of the landslide as the PLL site location. An AML *Value Query* that allows inferring the *location* attribute is:

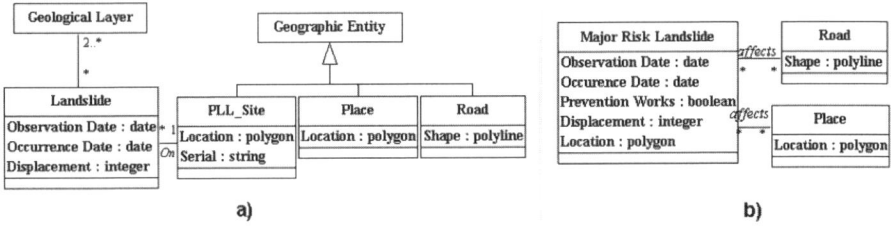

a)   b)

Fig. 4. Example of a base data schema (a) with a derived "regional planer" viewpoint (b)

```
this.Location = this! On.PLL Site.Location
```

Another transformation consists in renaming the main class (*Landslide*) into *Major Risk Landslide*, and to include in it only objects of interest to regional planners, i.e. landslides which present a major risk for the population or the infrastructures. An AML *Objects Query* that allows retrieving the objects member of the new class is:

```
this.objects = set (l in Landslide: (exists v in Place:
distance (l.Location, v.Location) < 500 or exists r in
Road: distance (l.Location, r.Shape) < 100) and
l.Displacement > 500)
```

This query allows presenting to regional planners only the landslides bigger than a given threshold (a displacement superior to 500 kg.) and which, by their position close to inhabited places or roads (within 500 m. of inhabited places or within 100 m. of roads), might present a potential danger.

It is important to notice that this viewpoint derivation mechanism is usable only for data viewpoint visualization, because only the mapping from the base data schema to the derived one is provided. In order allow end users to update the data through a certain viewpoint, the inverse mapping should be provided by the designers. However, for some situations, considerable efforts are required from the designers in order to provide two way mappings and to insure that the mappings are consistent That is why we are currently working at implementing automatic viewpoint management mechanisms into the AROM-ST language, which would automatically insure data consistency, simplifying the design work and allowing to remain at a conceptual level.

5 Presentation Adaptation

In order to create interactive presentations, it is necessary to follow two successive steps. The first step defines the static structure of the presentation, the elements that compose the presentation and the manner in which they aggregate into more complex elements (for example, the layers form frames and the frames form visualizations, etc.). The second step defines the behavior of the various structural elements (changes that they undergo) during the interaction with the user.

Presentations in ASTIS encompass four aggregation levels (see Fig. 5). ASTIS transforms each basic informational unit (i.e. an attribute in an object) in a visual element, displayable on the screen. This implies a conversion of this attribute from the storage format (class *Attribute*) towards the display format, which can be graphical or textual (class *Visual Element*), by applying a series of style options (class *Style*).

A *layer* defines a set of visual elements belonging to objects of the same class. It is the equivalent of a set of attribute values from the data (for instance, the set of village contours). Multiple layers can be defined from the same class of objects, as different attributes can be used to represent the same object. For instance, a class of geographic objects (villages) can be represented on a map by different layers: one layer could contain the geometrical shape of the villages, while another layer could contain the center points of the same villages. A *frame* (map, time diagram or table) is composed of layers. Several synchronized frames can be aggregated to form a visualization.

In order to allow users to grasp all the aspects of information related to natural risks, ASTIS allows visualizing data according to three viewpoints simultaneously:

– Spatial frames (see class *Spatial Frame* in Fig. 5) allow visualizing data in their spatial context and executing spatial queries;
– Temporal frames (see class *Temporal Frame* in Fig. 5) allow visualizing data in their temporal context and to carry out temporal queries;
– Informational frames (see class *Thematic Frame* in Fig. 5) allow visualizing information in a textual form and to carry out queries in this form.

The interface of ASTIS is interactive, allowing the user to execute visual queries and to view their results. Each action of the user on one of the frames of the application triggers two successive actions of the system: *i*) the query is interpreted and carried out and *ii*) the results are displayed on all the frames in a synchronized way.

Two complementary mechanisms are available for making visual queries:

1. **Masking** (hiding) objects that are not significant, by filtering mechanisms on spatial, temporal or thematic criteria. The main idea of masking is that, at any given moment, an object must either be visible on all the frames of the application or invisible on all of them. If the user masks an object on one of the frames, the system masks all the aspects of the object in all the frames. For example, when a user moves the map so that the geographical position of a landslide is not visible any more, the point that represents this event on a temporal frame disappears, as well as the thematic information (displayed on a thematic frame) describing it.

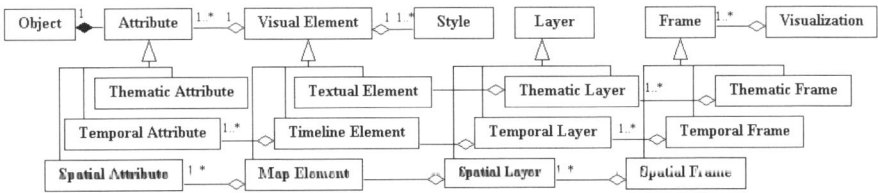

Fig. 5. The presentation model of ASTIS

2. **Highlighting** significant objects by selecting the objects that are in conformity with spatial, temporal or thematic criteria. Object selection (brushing) is an interaction technique used for a long time in the field of spatial interfaces [3], and for multi-view spatio-temporal interfaces [6]. Selection allows obtaining detailed information on some graphic objects. By employing it in conjunction with masking, it is possible to explore data in a very effective way: filtering allows quickly (but also with coarser grain) excluding irrelevant objects and the selection allows refining the query by including only the relevant objects.

Although complementary and similar, the two mechanisms are not identical in their functioning: masking affects all the layers composing a frame. Highlighting can be applied only to one layer at a time, while the user has the possibility to change the active layer for selection. Each interaction of the user with the interface of ASTIS is interpreted as a query: the contents of the frames are refreshed permanently in answer to the actions of the user. The queries are formulated using the algebraic modeling language (AML) of AROM-ST. The general form of a query for each data layer is:

```
set (object in Class: SR and TR and ThR)
```

The expression of the query for displaying a layer is a filter defined upon three restrictions: a spatial restriction (*SR*), a temporal restriction (*TR*) and a thematic restriction (*ThR*).All types of frames allow object selection:

```
SR (or TR or ThR) = object member SelectedSet
```

Spatial and temporal frames allow masking instances by *zooming* or *panning*, displaying only the instances situated in the visible spatial zone (*viewport*):

```
SR = intersects (object.Geometry, viewport)
```

Thematic frames allow combining sorting with the selection of objects (a technique called *focusing* [2], [7]). The user can sort instances displayed in a thematic frame by increasing or decreasing order of attribute values, and then they can use selection on the sorted set of instances. It is thus possible, for instance, to sort landslides by size and to select only critical ones (those which exceed a certain threshold).

After the execution of the query, the system must display its results on all the frames of the application. In order to allow the users to visualize sets of multidimensional and complex information, ASTIS uses synchronization mechanisms (or *linking* [2], [6]). Our use of synchronization in ASTIS is based on the idea that all the graphic elements, describing aspects of the same object in different views, must be coherent at all times. The visual state (visibility, invisibility, activation or deactivation) is a feature describing the object as a whole and must be reflected on all the visual elements describing it. Synchronization allows following and visualizing the various aspects of the same complex object through several views. Spatio-temporal objects can thus be known in their thematic aspects as well as in their spatial and temporal aspects.

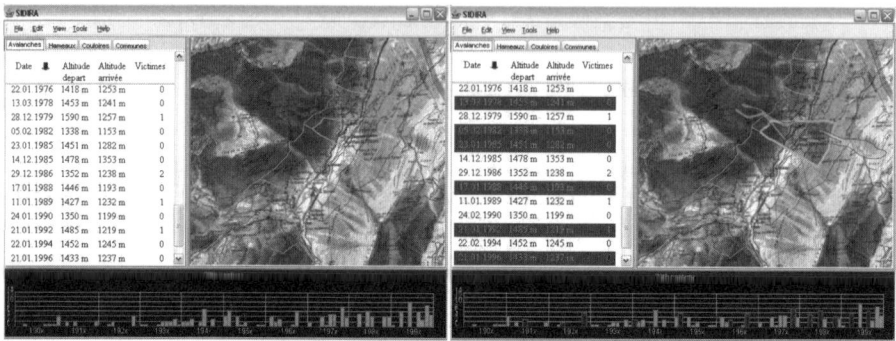

Fig. 6. Example of synchronization in ASTIS

In ASTIS, the synchronization mechanism has been extended in order to allow the visualization of relations between objects. The visual state is transmissible between objects that are related in the knowledge base. By selecting an object in a view, for example, it is possible to find the objects linked to it in the other views, these latter being highlighted. Let us take the example of a visualization dedicated to avalanche risks (see Fig. 6). By selecting the contour of a village, one can see in the spatial frame all the avalanche sites which intersect it. This, in turn, allows seeing (in the spatial and temporal frames) the avalanches that have affected these sites (and thus, the village).

Viewpoints may influence not only the data to be presented to users, but also the way the data are presented to them [15]. In order to build customized presentations, the ISNR designers must instantiate the model presented in Fig.7, which is a simplified version of the presentation model of ASTIS. Once the model created, ASTIS reads it and generates a presentation in conformity with the model.

The model allows designers to create layers by specifying for each layer the class of objects (attribute *Class* in class *Layer*) and the specific attribute which composes it. The attribute is of spatial type for the spatial layers (attribute *Spatial Attribute*) and of temporal type for the temporal layers (attribute *Temporal Attribute*). In order to create thematic maps and thematic time diagrams, the designer must also specify the thematic attribute (attribute *Attribute* of the class *Thematic Spatial* and, respectively, *Thematic Temporal*). The *Link* class allows synchronizing the data layers, the designer must specify the association or the succession of associations (attribute *Association Path*) connecting the classes of the two layers in the data model. This is necessary only for the visualization of the relations between objects, the synchronization of the various aspects of the same objects being managed automatically by the system.

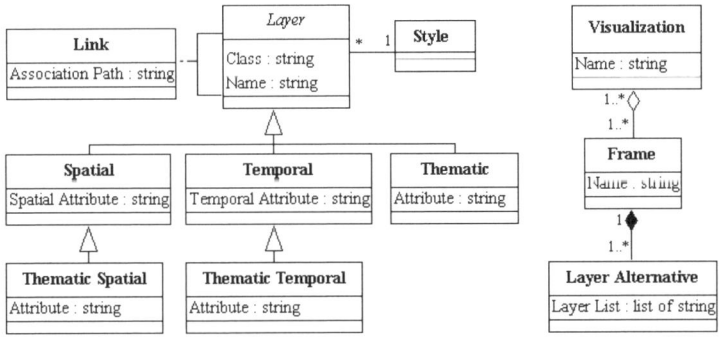

Fig. 7. The presentation model to be instantiated by the designers

In order to complete the specification of the presentation, the decomposition of visualizations into frames and of the frames into layers must be described. The designer may describe alternate layers for the creation of the frames. This allows, for example, creating a map which contains two layers: *i*) a spatial layer with the outlines of the communes of a certain area and *ii*) a thematic spatial layer which is a representation by means of proportional circles of the population of the communes, a representation by proportional circles of the GNP of the communes or a chloropleth representation of the GNP/population ratio of the communes (see the example in Fig. 8a).

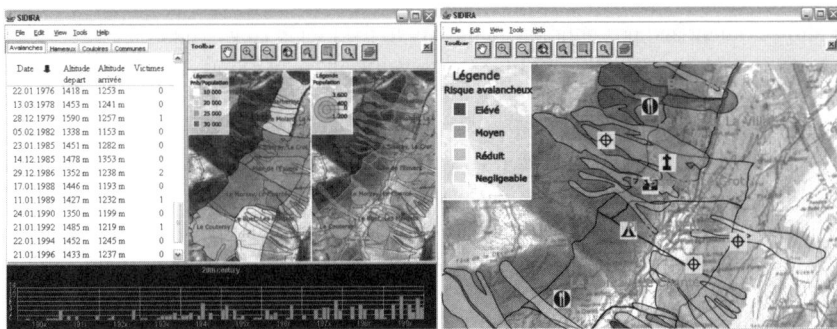

Fig. 8. Comparison of two viewpoints for the same application

We now give a small example of how presentations can be customized for a certain viewpoint. Let us consider the design of presentations for a "regional planner" and a "tourist" viewpoint in an application dedicated to the study of avalanche risks (see Fig. 8). Some content differences can be noticed between the two visualizations the regional planner viewpoint includes data such as villages GDP, population, risk alleviating investments, etc., while the "tourist" viewpoint contains data related to sights, restaurants and accommodation. Leaving aside the context differences, the structure of the presentations also differs. In designing a presentation for regional planners, the temporal perspective is very important. The likeliness of reoccurrence of avalanches is very important for making informed regional planning decisions, so a presentation for a "regional planner" viewpoint should include a temporal frame in order to allow temporal navigation. It should also include, most likely, maps and time diagrams with layers based on spatial and temporal analysis methods. These presentations should help them make decisions for alleviating risks as much as possible with expenses as little as possible. For a "tourist" viewpoint on the same application, other types of visualizations could be suitable. As tourists are interested the very short term (one or two days) forecast of avalanche risks, a temporal frame might not be necessary. Further on, the presentation could rely more on visual symbols than on displaying textual information. Last, but not least, the amount of visual detail displayed by the two presentations differs, as it is considered that regional planners have a high level of expertise with spatial applications.

6 Conclusion and Future Developments

We presented in this paper an environment for the design and the generation of spatio-temporal information systems dedicated to the study and management of natural hazards risks, called ASTIS. ASTIS proposes an approach allowing ISNR designers to generate spatio-temporal information applications only by creating conceptual models. By creating appropriate models, designers can configure each of the three modules of the architecture (the data, presentation and the adaptation modules) in order to obtain application that allow content and presentation adaptivity to the user's viewpoints.

Our approach is currently used for the generation of an ISNR dedicated to the multidisciplinary study of several classes of natural hazard risks (landslides, avalanches and floods) called SIRHEN. The SIRHEN project gathers scientists (geographers, geologists, historians and computer scientists), decision makers and regional planners and aims at a better knowledge about natural hazard risks, in order to allow for better decisions of land development.

A first direction of development of our approach was already mentioned, aiming at simplifying the work of the designers for viewpoint management, by including automatic viewpoint management mechanisms in the object-based knowledge representation system AROM, used by our framework.

Other current researches are aimed at improving the adaptive mechanisms. In the current state of development of the application, only features of the user (viewpoint, expertise level, etc.) are taken into account for the adaptation. Being given the development of ubiquitous computing and the fact that field work is essential in the natural hazard risk management, we plan to add to ASTIS automatic (without the need for an intervention from the designer or from end-users) support for adaptivity to features related to the user context. These features include the access devices of the users, their location, time and their current activity.

Bibliography

1. H. Anegg, H. Kunczier, E. Michlmayr, G. Pospischil, M. Umlauft: LoL@: Designing a Location Based UMTS Application, ÖVE-Verbandszeitschrift e&i, Springer, Heidelberg, Germany (2002)
2. Buja, A., McDonald, J. A., Michalak, J., Stuetzle, W.: Interactive data visualization using focusing and linking. Proc. of the 2nd Conference on Visualization '91 (1991)
3. Dykes, J.: Exploring spatial data representation with dynamic graphics, Computers & Geosciences, 23(4). (1997) 345-370
4. Kobsa, A., J. Koenemann and W. Pohl: Personalized Hypermedia Presentation Techniques for Improving Online Customer Relationships. *The Knowledge Engineering Review* 16(2), (2001) 111-155
5. MacEachren, A.M.: VISUALIZATION - Cartography for the 21st century. 7th Annual Conference of Polish Spatial Information Association. Warsaw, Poland (1998) 287-296
6. MacEachren, A.M., Boscoe, F.P., Haug, D., Pickle, L. W.: Geographic Visualization: Designing Manipulable Maps for Exploring Temporally Varying Georeferenced Statistics In Proc. Infovis, (1998)
7. MacEachren, A. M., Howard, D., von Wyss, M., Askov, D., Taormino, T.: Visualizing the health of Chesapeake Bay: An uncertain endeavor. Proc. GIS/LIS '93 Minneapolis, 2-4 Nov., (1993) 449-458
8. Moisuc, B., Davoine, P.-A., Gensel, J., Martin, H.: Design of Spatio-Temporal Information Systems for Natural Risk Management with an Object-Based Knowledge Representation Approach, Geomatica, Vol. 59, No. 4 (2005)
9. Monmonier, M.: Geographic brushing: Enhancing exploratory analysis of the scatterplot matrix, Geographical Analysis, 21(1) (1989) 81-84
10. Nivala, A-M., Sarjakoski, L.T. : Preventing Interruptions in Mobile Map Reading Process by Personalisation. The 3rd Workshop on "HCI in Mobile Guides", in adjunction to: *MobileHCI'04, 6th International Conference on Human Computer Interaction with Mobile Devices and Services*, September 13-16, 2004, Glasgow, Scotland, 6 p.
11. Page, M., Gensel, J., Capponi, C., Bruley, C., Genoud, P., Ziebelin, D., Bardou, D., Dupierris, V.: A New Approach to Object-Based Knowledge Representation: the AROM System, Lecture Notes in Artificial Intelligence (2001)
12. Parent, C., Spaccapietra, S., Zimanyi, E.,: Spatio-Temporal Conceptual Models: Data Structures + Spatial + Time, Proc. ACM GIS: 26 – 33 (1999)
13. Schwinger, W., Grün, Ch., Pröll, B., Retschitzegger, W. and Schauerhuber. A., Context-awareness in Mobile Tourism Guides - A Comprehensive Survey, Technical Report (2005)
14. Sester, M. and Brenner, C.: Continuous Generalization for Fast and Smooth Visualization on Small Displays, *International Archives of Photogrammetry, Remote Sensing and Spatial Information Sciences*, XXXV (B4:IV): (2004) 1293-1298
15. Teraoka, T. and Maruyama, M.: Adaptive information visualization based on the user's multiple viewpoints-interactive 3D visualization of the WWW. In Proc. INFOVIS (1997)
16. Tufte, E.: The Visual Display of Quantitative Information. Cheshire, Graphics Press (1983)
17. Woodruff, A., Landay, J., Stonebraker, M.: Constant Information Density in Zoomable Interfaces. Proc AVI'98, L'Aquila, Italy (1998)
18. Zipf, A. User-Adaptive Maps for Location-Based Services (LBS) for Tourism. Proc. ENTER 2002, Springer, Heidelberg (2002)

A Contextual Approach for the Development of GIS: Application to Maritime Navigation

Mathieu Petit, Cyril Ray, and Christophe Claramunt

Naval Academy Research Institute, 29240, Brest Naval BP 600, France
{petit, ray, claramunt}@ecole-navale.fr

Abstract. The research presented in this paper introduces the principles of a multi-dimensional contextual approach for adaptive GIS. The framework makes the difference beetween the user, geographical and device contexts. The geographical context is modelled according to the location of the user, the region of interest, the extent of the region covered by the diffusion of the data, and the place where the information is processed. This characterization allows for the study of the different contextual configurations, and their impact on the design of mobile services. The framework is applied to maritime navigation.

1 Introduction

Technological advances observed over the past few years have favoured the emergence of mobile computing as a novel trend for information diffusion. On the one hand, this opens many opportunities for fullfilling the large range of user needs, but on the other hand current mobile systems often suffer from many limitations such as a lack of connectivity, poor interface design and a dramatic lack of memory and computing power.

Nowadays, recent progress made in energy consumption and computing power, display and memory sizes, and interactive tools have reached a threshold that allows GIS research and development communities to explore novel mobile applications and interfaces. One of the most important technological trends is the progressive integration of different communication techniques that allow mobile GIS and appliances to integrate, process and exchange data using wireless communications. Another trend is the integration of geolocalisation systems within mobile appliances that deliver geographical information for embedded GIS applications [1][7]. These technological advances offer new opportunities for the design and development of mobile GIS.

The research presented in this paper introduces a context-aware mobile GIS that integrates adaptive interaction techniques. We define an adaptive GIS as a generic GIS that can be automatically adapted according to several contexts defined by (1) the properties and location of the geographical data manipulated, (2) the underlying categories that reflect different user profiles and (3) the characteristics of the computing systems, supporting web and wireless techniques. This classification has been inspired by a previous work done by Calvary *et al.*

J.D. Carswell and T. Tezuka (Eds.): W2GIS 2006, LNCS 4295, pp. 158–169, 2006.

[2]. These contexts cover the components of the diffusion of geographical data in wireless environments. The dimensions identified are of different nature as they involve data, computing processes and interfaces, and categories of users.

These requirements are not new when studied individually, but less considered as a whole. For instance, previous work in the field of adaptive GIS introduces a technology-driven approach for an hardware-based interaction medium [6]. Adaptation of an open GIS layer descriptor to specific user needs and contexts have been also studied in [11]. A context-sensitive model for mobile cartography that emphasizes different levels of data adaptation and presentation have been proposed in [9]. In a previous work, we introduced an architecture and real-time services for the diffusion of maritime geographical information, at different levels from the global monitoring of the maritime traffic of a given area [4], to individual services on request [5].

In order to consider the problem from a global point of view, we introduce a research whose objective is to develop an integrated contextual-based architecture that considers these different factors and interrelationships. The framework is developed and applied to maritime navigation, an emerging field of GIS that combines mobility and distributed services. The remainder of this paper is organised as follows. Section 2 introduces our modelling of a context-aware GIS. Section 3 presents a preliminary application of our framework to maritime navigation. Finally section 4 concludes the paper and draws some perspectives.

2   A Generic Model for Adaptive GIS

The main idea behind an adaptive GIS relies in its capacity to automatically derive its content and interface from a changing environment. This assertion raises the following issue: the contextual dimensions should be clearly identified, and supported by flexible and dynamic algorithms, and adaptive computing processes that support interactions with the users. This constitutes a three-level modelling environment (Fig. 1) whose dimensions can be characterized by the user context (i.e. who), the geographical context (i.e. what) and the appliance context (i.e. how).

The relationships between these dimensions constitute the target of an adaptive process and the subject of our modelling approach. On the software and interaction sides, the system should deliver the geographical data the user may interact with. Geographical data is delivered by a generic communication and integration layer whose role is to aggregate different data flows from either real-time infrastructures, or previously stored geographical information, and to provide an homogeneous source to the adaptive GIS. Another objective of this layer is the internal data storage, that is, monitoring and storing incoming data in order to replay sequences of localized events and scenes that present an interest to the user. As this layer is context independent and constitutes the input of the adaptive GIS, data replays should be adapted to various targets. For instance, a whole geographical scene can be simulated off-line for debriefing or learning

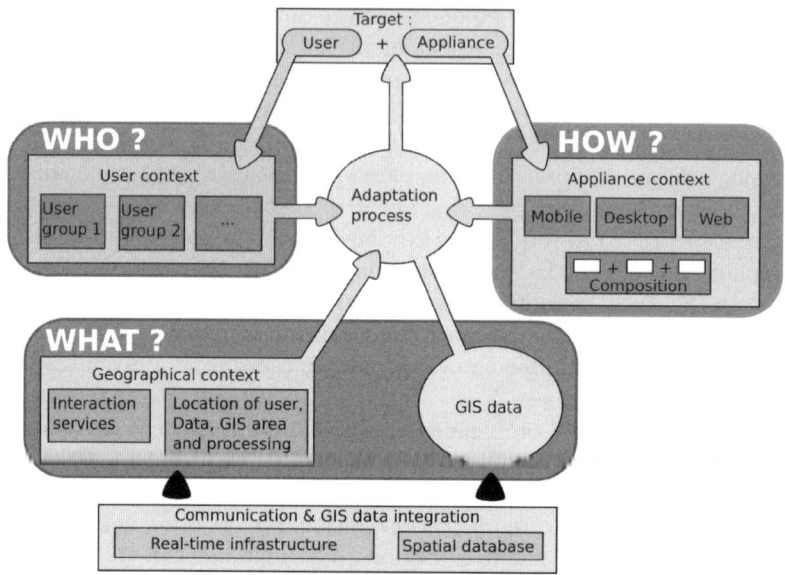

Fig. 1. Context-aware architecture of an adaptive GIS

purposes. A real-time scene can be also enriched with previous scenes to compare current with past situations [3]. This shows that an adaptive GIS constitutes an intermediate layer between data integration and presentation. Within an adaptive GIS, geographical data should be presented to the user with a specific attention to her/his characteristics, and the current context. An adaptation process should integrate elements from different contexts, and apply sorting algorithms to determine the relevant data, and present them in an automatically generated human-computer interface. Taking into account the contextual environment, a mobile GIS should improve the usability and usage of an application. Each contextual class owns its proper means to acquire, characterize and store contextual elements.

The user context reflects the way individual users are sorted into groups of similar behaviours according to the properties of the data usually requested, and the user interface usage. The appliance context characterises the internal specifications (e.g. data transmission speeds and volumes, interface memory), output capabilities (e.g. display size and resolution) and input capabilities (e.g. mouse, touch screen, keyboard). These are the main elements of the appliance context that have to be taken into account by the adaptive process. Appliances are organised into groups of similar capabilities and can be even composed of several devices to support groupwork.

These contextual parameters also constrain the design of the user's interface, that is, the choice and placement of widgets, the user application dialog and the functionalities proposed to the user [10].

2.1   Geographical Context

Geographical data is usually presented to the users by derived views which are available at a given time and space. Each geographical data view is generated by an appropriate service that diffuse its data to different users. As a collection of views, an adaptive GIS is also distributed in space. With respect to the spatial dimension, several locations of significance have been introduced to characterize the geographical context of the execution of a given service [8]:

- U: the location of the user and the interface from where the user obtains GIS-based information. Most of the time, U can be assumed as punctual as the user is likely to be in front of its appliance;
- D: the region of the data is available;
- P: the region where the data is processed;
- S: the region of interest of the GIS project.

Unlike common GIS where U, P and D are static and direectly integrated within the user desktop, in a mobile GIS, locations are distributed and dynamic over space and time. These four orthogonal locations constitute together a multi-dimensional space whose different configurations can be explored as they are likely to impact the way a given mobile user interacts with geographical data. Our intention is to characterize the range of possible configurations, and to which extent these influence and constrain the services delivered by an adaptive GIS. The range of possible geographical contexts is given by the combination of intersecting and non-intersecting binary relationships between these regions (cf. sample given by Tab. 1).

Table 1. Example of geographical context

When approaching a harbour, a tanker may be guided by an auxiliary vessel, either from the harbour's authorities or from the tanker itself (Fig. 2(a)). The data from the region of interest S is sent over a region D covering the surroundings of the harbour. The tanker processes approaching operations according to the delivered geographical data. Those processing results are made available into the region P around the tanker. The user U on the auxiliary vessel may be able to interact with the data through the tanker processing capabilities. The geographical context and their interactions are given by: $U \cap D \neq \emptyset, U \cap P \neq \emptyset, D \cap P \neq \emptyset$ and $S \cap D \neq \emptyset$. As P and U are linked to the mobile vessel and tanker, the geographical context is likely to change. If the vessel is moving away from the tanker, then at some point $U \cap P = \emptyset$ and the user will not be able to receive additional data processing results.

In order to represent the possible configurations, a tabular notation is introduced (Fig. 2(b)). Per convention, a black cell represents a non-empty intersection, while a white cell denotes an empty intersection between the location of significance. The combination of these binary spatial relationships generates the

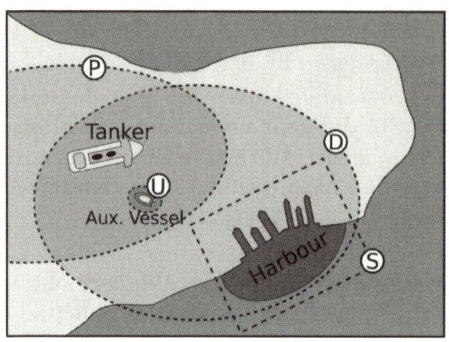

(a) Context-based geographical distribution

(b) Corresponding geographical
and contextual relationships:
$U \cap D \neq \emptyset, U \cap P \neq \emptyset, D \cap P \neq \emptyset$
and $S \cap D \neq \emptyset$

Fig. 2. Example of geographical context characterization

complete set of 64 orthogonal contextual configurations presented in Fig. 3. The
primitive contextual configurations of a binary relation can be summarized by
the following roles:

- $U \cap S \neq \emptyset$: the user is an *actor* in the region of interest;
- $D \cap S \neq \emptyset$: the data is diffused in a part of the region of interest (i.e. *local data diffusion*);
- $P \cap S \neq \emptyset$: the process of a service is in a part of the region of interest (i.e. *local processing*);
- $U \cap D \neq \emptyset$: the user receives some geographical data as she/he is located in the region of data diffusion (i.e. *data reception*);
- $P \cap D \neq \emptyset$: the processing service is available in a part of the region of data diffusion (i.e. *service at disposal*);
- $P \cap U \neq \emptyset$: the processing service is available to the user (i.e. *service accessed*).

The minimum requirements for a service to generate a geographical data view
are the simultaneous presence of a service *at disposal* and *accessed* by the user.
This implies at minimum $D \cap P \neq \emptyset$ and $U \cap P \neq \emptyset$; 16 combinations out
of 64 contextual configurations fullfill this constraint (Fig. 3 - dark grey cells).
However, whenever the location of the user intersects other regions (Fig. 3 - light
grey cells), these cases are of interest, as an adaptation should occur even when
no service is available. For instance, 8 combinations provide no interactions with
the user (Fig. 3 - first line) and cannot be considered by an adaptive process.
The combination $[c, 4]$ illustrates the desktop use of a non-distributed GIS; all
the other combinations denote a certain degree of distribution by the GIS service
components.

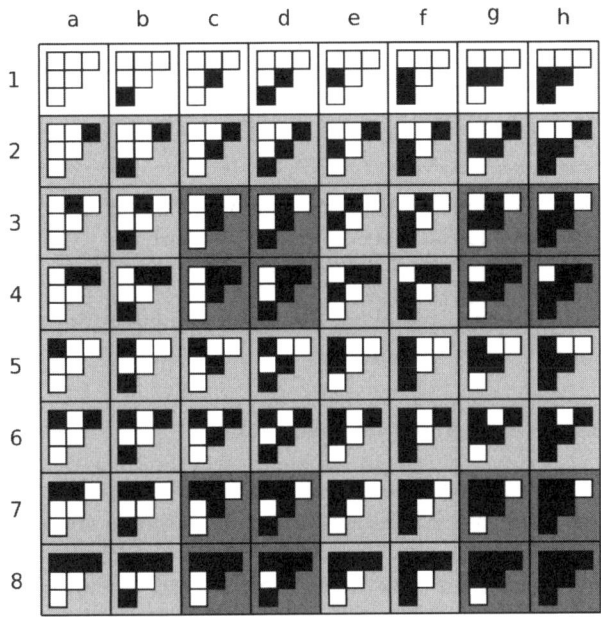

Fig. 3. Possible combinations of the regions of significance into 64 contextual configurations

2.2 Adaptive Process

An adaptive process considers the different contextual parameters, and produces a target-adapted application. The adaptation occurs on the two main components of the adaptive GIS:

- The container, that is, the user interface, by taking into account the user and appliance contexts;
- The content, that is, the geographical data views provided by the services, by considering the characteristics of the geographical context.

The adaptation should be performed at execution time, whenever one of the different contexts change. The user's context is likely to change when the user's behaviour evolves regarding her/his usage of the GIS functionalities and interface. The appliance context should change when the hardware capabilities are modified. The geographical context triggers some adaptive processes when moving from one region to another.

Considering the example proposed in Tab. 1, the user on the auxiliary vessel passes through a series of context cases during the guidance of the tanker to the harbour. Each step corresponds to an adaptation made of a progressive geographical context enrichment. Then a service is made available and is accessed when the data diffusion and processing regions intersect the user location. Fig. 4 illustrates a possible sequence of contextual configurations changes. Each role within a configuration involves a specific adaptation of the view provided by the service. When

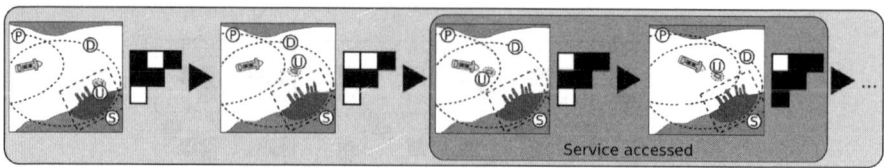

Fig. 4. Example of geographical context evolution : The auxiliary vessel leaves the harbour towards the tanker and drives it back to the harbour

the value of another binary spatial relationship is modified, the roles are updated before re-calculation of the view.

3   Maritime Navigation: Case Study

Our experimental framework has been tested in the context of an international sailing race. This event has a large audience, and requires appropriate solutions to diffuse real-time information, from the coastal maritime area to the users located in the ground. This generates different needs in term of geographical information usage and appliance. The experimental prototype is composed of two parts: a wireless network and an experimental adaptive GIS (Fig. 5).

Ships locations during the race are acquired through a real-time infrastructure. The implementation of the geographical context into different views and an appliance context divided into several classes of devices are considered by the adaptive process. The user context is modelled by a generic user group.

3.1   Communication and Data Integration

An important aspect of the adaptation process is its ability to integrate real-time geolocalisation information that delivers GIS data and influences the geographical context and related services. A localisation system has been developed and allows for real-time reception of ship's positions and a continuous video stream of the race (Fig. 6). locations are provided by an embedded system available on ships. This system includes a GPS, a configurable modem, a VHF transmitter and fulfills several constraints: light weight (less than a kg), long range (5 to 10 km), high autonomy (8 to 10 hours). This module collects and diffuses the real-time locations of the ships to the ground station (Fig. 6-(d)). The transmission to the ground station is a VHF communication (Fig. 6-(a)) based on APRS frames (Automatic Position Reporting System). The ground station is composed by a VHF receiver and a VHF-to-WiFi bridge that broadcasts real-time data to a given area (Fig. 6-(c_2)). Mobile end-users (Fig. 6-(f)) located in this broadcast area can access ships' data, whatever the form of their appliance.

A general drawback of coastal sailing races is the lack of visibility on the ground, because of the distance to the coast. As real-time positions are crucial, video streams are provided and offer a concrete service of the geographical data. The installed video system presented in Fig. 6-(e) broadcasts a video stream of the race (Fig. 6-(b)) to a given WiFi deserved region (Fig. 6-(c_1)).

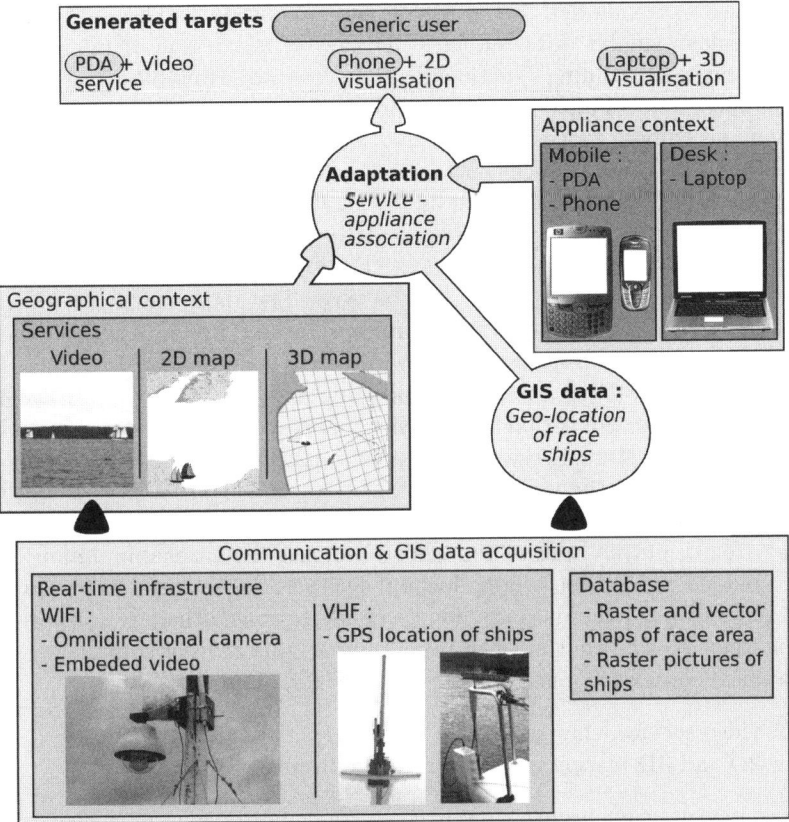

Fig. 5. Experimental implementation of an adaptive GIS model

3.2   Services and Geographical Context

Different geographical data views, each associated to a particular service, are presented to the user. The *"2D mapping"* service delivers ships location information. Different levels of zoom are automatically computed several times

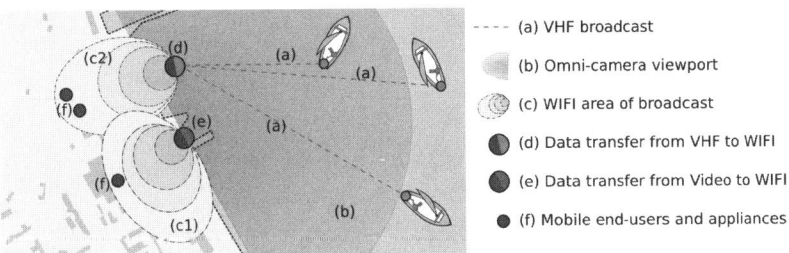

Fig. 6. Real-time communication infrastructure

per minute to provide detailed views on the race activity. The *"3D mapping"* service displays similar data but in a 3D view of the region of interest. Basic displacement and zooming functions in the scene are available using keyboard combinations. The *"Video"* service provides a real-time view of the race region. Zooming and camera movements are also allowed.

Each of these three services encompass a geographical context composed of the regions of significance, denoted as $\{U, D, P, S\}_{Video}$, $\{U, D, P, S\}_{2D}$, $\{U, D, P, S\}_{3D}$ for the user, data, processing and site regions, respectively for *"Video"*, *"2D mapping"*, *"3D mapping"* services (Fig. 7). As the region of interest is always the racing area, $S_{Video} = S_{2D} = S_{3D}$. The data server for D_{2D} and D_{3D} is a computer that collects and stores current and previous ship's locations and spreads the data in a 200 meters wide circular area, that is $D_{3D} = D_{2D}$. The data server for D_{Video} and the processing unit for P_{Video} are at the same location (i.e. image acquisition and video signal compression are done together), the resulting video stream is sent wireless over a 200 meters ellipsoid region. P_{3D} and P_{2D} are done by the user appliance, at the same location as U_{2D} and U_{3D}, respectively. The processing region is only local to the appliance. The service view at U_{Video} output videos and allows for the distant manipulation of the camera but no processing is done locally. The users are mobile over space and may intersect or not the data or the processing regions; others regions are fixed (except when $P = U$, that is when processing is done on user's appliance). Two geographical context changes are identified (Fig.3):

- For video service, change from $[c, 4]$ to $[c, 1]$
- For 2D and 3D mapping services, change from $[c, 4]$ to $[a, 3]$

These contextual changes are supported by immediate transitions, that is, with no intermediary contextual configurations. However, and in order to ensure a progressive enrichment of the view, intermediate contextual configurations are inserted between initial and final configurations. A given sequence is constructed where two successive configurations differs from one and only one role (Fig. 8). For the video service, these sequences can be represented by $[c, 4] \rightleftarrows [c, 3] \rightleftarrows [c, 1]$ or

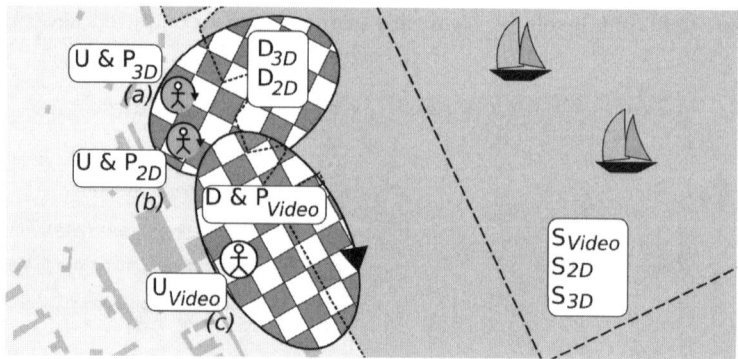

Fig. 7. Geographical context applied to maritime navigation

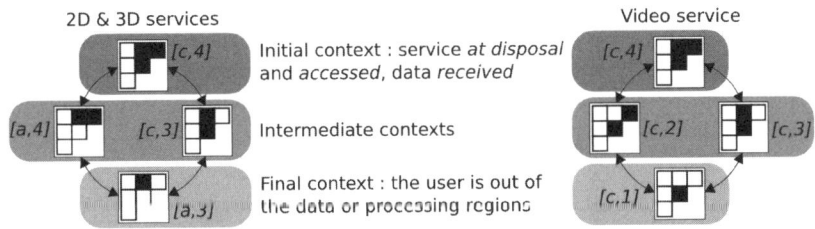

Fig. 8. Transition between geographical contexts

$[c,4] \rightleftarrows [c,2] \rightleftarrows [c,1]$; whereas for the 2D and 3D mapping services, the sequences are $[c,4] \rightleftarrows [a,4] \rightleftarrows [a,3]$ or $[c,4] \rightleftarrows [c,3] \rightleftarrows [a,3]$. Fig. 7 presents the case of the users (a) and (b) interacting with two appliances that run the "2D mapping" and "3D mapping" services, respectively. The processing is done at the user's location. (a) and (b) are located in a region where 2D and 3D data are transmitted. When the users leave their region, the intermediate context $[a,4]$ or $[c,3]$ occurs simultaneously with the final context $[a,3]$. A third user (c) is located in a region where the "video" service is available. As P_{video} and D_{video} are the same region, when (c) leaves the region, the intermediate contexts $[c,2]$ or $[c,3]$ occur simultaneously with the final context $[c,1]$. The "Video" service view is adapted when contextual configuration switches to $[c,4]$ or $[c,1]$ as follows (Fig. 9(a)):

- $[c,4]$: the user is located in the region of video streaming. The adaptive GIS is enriched with a new view.
- $[c,1]$: as the user is outside the video broadcasting region, the service is not available anymore. On the appliance, no views are provided. The adaptive GIS is waiting for a new service to come up.

The "2D mapping" and "3D mapping" services share the same geographical contexts, adaptation processes and changes at the interface level occur as follows (Fig. 9(b)):

- $[c,4]$: the user and the processing services are located in the region of geographical data diffusion, a view showing either a 3D or 2D map is displayed on the device.
- $[a,3]$: the 2D and 3D services run on the user appliance but no data is provided. The service does not update the data anymore. On the appliance side, the view shows the last data received and informs the user that no more data are available for transmission.

The appliance context is also taken into account by the adaptive process. A mapping between the generated view and a particular appliance is applied. The view generated by a *"2D mapping"* service is displayed on a Java-enabled mobile phone as this service doesn't requires substantial graphical capabilities and as it is adapted to small displays. The *"3D mapping"* view is adapted to the desktop PC with sufficient computing power and memory. The *"Video"* view is displayed on a mobile PDA. Regarding the appliance a user is interacting with, the services

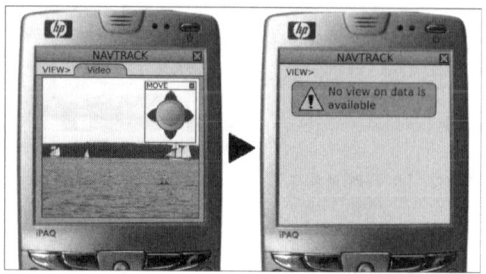 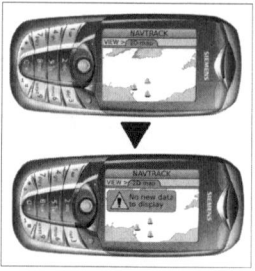

(a) Adaptation to contexts $[c, 4]$ and $[c, 1]$ of the video service

(b) Adaptation to contexts $[c, 4]$ and $[a, 3]$ of the 2D mapping service

Fig. 9. User-interface adaptation triggered by geographical context changes

are made available or not. For example, if a user with a PDA gets in the *"3D mapping"* broadcasting region, no new view can be provided as the associated service is only suitable for users acting on desktop PCs. Fig. 9 shows that the generic user interface is also adapted to the appliance features (e.g. screen size and orientation).

4   Conclusion and Future Work

Although users and appliances are essential components of mobile GISs, they are not always taken into account in the processing and display of geographical data. The research presented in this paper introduces a contextual-based modelling approach that considers users, appliances and geographical data as the core elements of an adaptive GIS. The model proposed identifies and characterises different elements that constitute the geographical context: user, data, process and region of interest. Integration of these elements allows for the identification of different configurations. This influences the way interactions between the user and GIS services should be adapted. This approach is applied to the context of maritime navigation and illustrated by a prototype that support the adaptation of the services proposed to the user regarding an evolving geographical context, and the appliance characteristics. Future works concern the integration of the user and appliance contexts and the modelling of evolving configurations using conceptual neighbourhood graphs.

Acknowledgement

The authors are gratefull to the Science Park Technopole Brest-Iroise for their participation to this work. Special thanks to Yves Vourc'h, Roland Lechallier, Mathieu Goury, Guillaume Mevel and Fabien Nicolas for their help in the implementation of the wireless infrastructure.

References

1. D.G. Abowd, G.D. Atkeson, J. Hong, S. Long, R. Kooper, and M. Pinkerton. Cyberguide : a mobile context-aware tour guide. In *Proceedings of ACM Wireless Networks*, pages 421–433. ACM press, 1997.
2. G. Calvary, J. Coutaz, D. Thevenin, Q. Limbourg, L. Bouillon, and J. Vanderdonckt. A unifying reference framework for multi-target user interfaces. *Interacting with Computers*, 15(3):289–308, 2003.
3. C. Claramunt, S. Fournier, X. Li, and E. Petchev. Real-time geographical information systems for its. In *Proceedings of the 5th IEEE International Conference on ITS Telecommunications*, pages 237–242, 2005.
4. G. Desvignes, G. Lucas de Couville, E. Peytchev, T. Devogele, S. Fournier, and C. Claramunt. The Share-loc project: a wap-based maritime location system. In B. Huang *et al.*, editor, *Proceedings of the 3rd International Conference on Web Information Systems Engineering*, pages 88–94. IEEE press, 2002.
5. J. Dubs and R. Kaufmann. Gestion et visualisation de données provenant d'un AIS. Technical report, Naval Academy Research Institute and Université de la Rochelle, 2006.
6. H. Hampe and V. Paelke. Adaptive methods for mobile applications. In *Proceedings of the 7th International Conference on Human Computer Interaction with Mobile Devices and Services (MobileHCI 2005)*, 2005.
7. B. Huang. Developing location-aware navigation guides that uses mobile geographic information systems. *Journal of Transportation Research Record*, 1879:108–113, 2005.
8. P.A. Longley, M.F. Goodchild, D.J. Maguire, and D.W. Rhind. *Geographical Information Systems and Sciences*. John Wiley and Sons, 2nd edition, 2005. 517 pages.
9. T. Reichenbacher. Adaptive methods for mobile cartography. In *Proceedings of the 21th International Cartographic Conference*, pages 1311–1322, 2003.
10. D. Thevenin. *Adaptation en Interaction Homme-Machine : le cas de la plasticité*. PhD thesis, Université Joseph Fourrier, Grenoble, 2001.
11. A. Zipf. Using styled layer descriptor (SLD) for the dynamic generation of user- and context-adaptative mobile maps - a technical framework. In K.J. Li and C. Vangenot, editors, *Proceedings of the 5th International Workshop on Web and Wireless GIS (W2GIS 2005)*, pages 183–193. Springer-Verlag, 2005.

P2P and Agent Service Based On-Line 3DGIS

Xi-Cheng Tan and Fu-Ling Bian

Research Center of Spatial Information and Digital Engineering , Wuhan University,
129 Luoyu Rode, Wuhan, China 430079
txcdhp2003@163.com

Abstract. Currently, the 3DGIS on the web can not meet the requirement of many applications because of the low network bandwidth, the
large amount data involved, and the need of high speed of data transportation. To resolve the problems, in this paper, the on-line 3DGIS
based on P2P agent distributed computing environment is proposed. We
designed the tile-based 3D terrain and the mechanism of the distribution and management of 3DGIS data on P2P agent environment. We also
proposed data searching and transmission algorithms for on-line 3DGIS.
Finally, the data security mechanism of on-line 3DGIS is presented.

Keywords: 3DGIS; P2P; Agent; Distributed Computing; Data Security;
DEM.

1 Introduction

The 3DGIS has been used in many areas, such as urban planning, telecommunication, real-estate marketing, environmental monitoring, flood prevention and
cure,weather simulation and military training. There are some on-line 3DGIS
applications, but they have many limitations. Because of the large amount data
of 3D terrain and texture, it is not easy to transfer them on the internet. So
the classical technique based on Client/Server(C/S) structure can not fulfill the
needs. At the same time, The Peer-to-Peer (P2P) and mobile Agent technology
have gotten great progresses. P2P has become an important distributed computing environment. With the P2P computing technique, one can exchange information and share data using different computing devices[1], and peers can share the
stored resources in a large-scale environment. The mobile agent is a software that
can move within the network and act on behalf of a user or another entity [2,3].
The mobile Agent can effectively reduce the burden and improve communication
efficiency of network. In this paper, we propose an agent based P2P system for
the better performance of on-line 3DGIS. The goals of this paper are:(1) to put
forward a P2P agent based 3DGIS infrastructure, which integrates the 3DGIS
with P2P and agent services; (2) to present a tile-based solution of the terrain
to facilitate the 3DGIS surfing;(3) to design a scheme of the distribution and
management of 3DGIS data on the P2P system; (4)to propose the mechanism
of terrain data processing of P2P Agent 3DGIS; (5) to propose the agent based
algorithms of searching and transporting 3DGIS data in P2P network; (6) to
present a scheme of the 3DGIS data security in the P2P agent system.

J.D. Carswell and T. Tezuka (Eds.): W2GIS 2006, LNCS 4295, pp. 170–179, 2006.
© Springer-Verlag Berlin Heidelberg 2006

2   Architecture of On-Line 3DGIS Based on P2P Agent System

The network structure of 3DGIS system based on P2P and mobile agent, called P2P Agent-3DGIS ,is shown in Fig.1. The system includes three kinds of peers: (1) original data source server peer, it's DS in Fig.1 ; (2)district server peers, they are Sn in Fig.1; (3) simple peers, they are Pi,j in Fig.1. The original data source server peer keeps the information of the district servers and it also keeps all the indexes of the 3DGIS data, because the peer belongs to provider of the 3DGIS service, and all the data is occupied by the provider exclusively. The district server peer keeps the data indexes of the simple peers. The simple peers keep the data and data indexes of themselves. In the system, when the peer's service starts, it will firstly request the original data source server and obtain the information of district servers, and then it will select the most appropriate district server to join in according to the network situation of the district servers.

District server peers are responsible for managing an arbitrary set of peers to be contacted [3], and providing the information to agents. The district server peer also plans the itineraries of mobile agent. The structure of P2P Agent 3DGIS component of district server peer is shown in Fig.2.

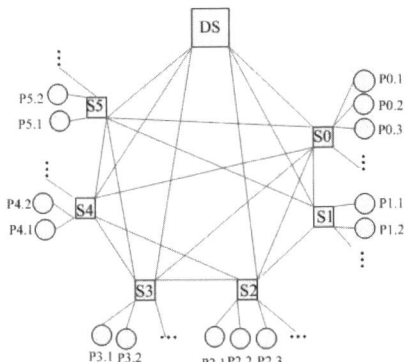

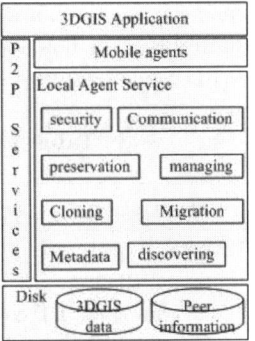

Fig. 1. Network structure of P2P Agent 3DGIS

Fig. 2. The structure of P2P Agent 3DGIS component

As shown in Fig.2, the component includes four parts: (1) P2P service ;(2) local agent service; (3) mobile agent service; (4) 3DGIS application .The data transportation takes place in P2P network. The data searching and moving itineraries of mobile agents are all planned by the local agent service. In order to join a P2P system a new peer initially contacts to the DS for the information of district servers, and then requests one of district sever peers. The peer provides the district server peer with its information about shared resources. The local agent run on the server peers fixedly, and its work is receiving the request of the mobile agent, and taking action differently according to different request, it can perform tasks from simple searching to complex logical session, it also completes the business such as communication with mobile agents, clone of agent and security etc. The mobile agents move through P2P network on the itineraries planned by the local agent and search information, at the same time it complete the communication of the peers. The 3DGIS application is constructed based on P2P agent service, and do the work such as data decoding, reading and displaying in 3 dimensions. The work such as data searching and transportation will be assigned to the P2P agent service.

3   The Construction of 3DGIS Terrain Tiles

As we know that because of the mass data of 3DGIS , we can not read all the data from server only one time. The terrain data including DEM data and remote sensing image data must be divided into many tiles called TTs(Terrain Tiles)[0]. The TT composes of DTs (DEM Tiles) and ITs (Image Tiles).The TT's size can be any value in theory, but it is limited by the capability of machines in practice. If the TT is divided too big, it may exceed the throughput of machine .If being divided too small, it will result in excessive activation of reading from and writing to the server. So the size of TT must be reasonable. Generally the amount of grid of the DT should be 2^n (n must be integer more than 2).In this paper, the used DEM data has a resolution of 20m, and resolution of SPOT remote sensing image is 2.5m. The DT has 16*16 square grids. Accordingly, the IT is a square block, it has a size of 16*(20/2.5), so the IT is a 128*128 matrix. The whole 3D Scene composes of 16*16 TTs .

Table 1. Notations used in this paper

Symbol	Description
DT	DEM tile
IT	Image tile
TT	Terrain tiles, TT composes of DTs (DEM Tiles) and ITs (Image Tiles)
TTIF	Terrain tiles index file ,it stores the TT tiles' indexes
DTDF	DEM tiles data file ,it stores the DTs data
ITDF	Image tiles data file ,it stores the ITs data

Each DT or IT has a unique indexical number, which is calculated according to the Dem data. The DEM layer should be divided into many foursquare tiles from top left corner. These foursquare tiles consist of the matrix of 17*17 points. Assume i is the row of the DT, and j is the column, the index of DT can be calculated according to equation(1):

$$\text{index of DT} = i \times \text{column of DEM} + j \qquad (1)$$

In the 3D scene, the application must always make the eyes close to the scene center, the distance between the eyes and the scene center will never exceed the size of TT , when the eyes move, the new TTs data will be read from the server, and the scene center will be changed accordingly. After the 3D terrain data are divided, they must be distributed on the peers of the P2P system.

4   The Distribution and Management of 3DGIS Data on P2P Agent System

In this system, the original data source server peer(DS) keeps all the TTs and their indexes, and it also keeps the information of the district servers. After a district server peer joins the P2P network, its information will be kept by DS, and the district server peer will manage the TTs data of simple peers that connecting with it, Fig.3(a) shows the data managing mechanism. In this way, when an agent is searching data, it is unnecessary to search the simple peers. The agent will find all the data of the peers only by searching the TTs indexes on the district server peer, and this will reduce the burden of the network remarkably. In this paper we take the data of a district of China as the testing data, there are about 200MB SPOT image data and more than 30MB DEM data. In the P2P system, these data can be distributed on many peers, the simple peers can store more than 50MB TTs data .So when the amount of Peer increase, the data will be distributed on more peers, these data may have much overlapped area, as shown in Fig.3(b), therefore the users can obtain more data sources, and the speed of reading data will be remarkably accelerated.

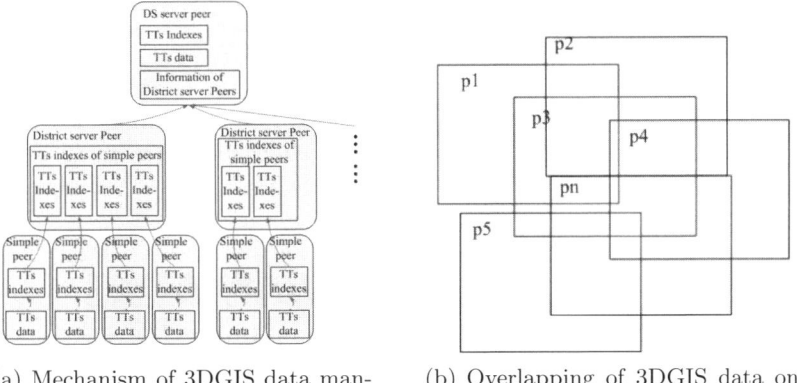

(a) Mechanism of 3DGIS data managing

(b) Overlapping of 3DGIS data on the peers

Fig. 3. Distribution and management of the 3DGIS data

All the 3DGIS data of a peer will be stored in the three files: DTDF(DEM tiles data file),ITDF(Image tiles data file) and TTIF(Terrain tiles index file). DTDF stores the DTs data, ITDF stores the ITs data, TTIF stores the TTs index. The stored data are the areas that the user of the peer is interested, for example, when the viewpoint rambles to point (x,y,z),the peer will read a block of terrain data center at (x,y) and store them. When the viewpoint moves on, the data of the area that the user rambled through will be stored in the three files too. When the 3DGIS application starts to run, it will read the TTs indexes from TTIF into RAM and send them to the district server peer. If there are newly added TTs data, the application will send these indexes to the district server peer intermittently. In this way, the district server will know the latest situation of the TTs of the simple peers. When the capacity of TTs data exceeds a limit, the application will tell the district server to delete the indexes of the TTs, which are far from the viewpoint, and then it will delete the TTs data from the files to save the space of the disk.

5   The Mechanism of Terrain Data Processing of P2P Agent 3DGIS

The data processing of P2P Agent 3DGIS application includes two parts of work: (1) scene data calculating and managing of 3DGIS application; (2) data searching and transportation based on P2P agent service.

Now, let's study the first part of work. Fig.4(a) shows the structure of the terrain data when the 3DGIS application runs into the scene at the first time. The whole terrain data are constituted by TTs. The small gray square in Fig.4(a) is the visible scene, and the tiles out of the small gray square is the TTs read beforehand. There are six TT-loops outside the small square. If the 3DGIS is Client/Server based application in LAN, it can ramble smoothly by reading only one TT-loop outside the scene beforehand, but on the internet it must read more TT-loops data outside the visible scene beforehand, because the latency of transportation of the on-line 3DGIS data is far more than that in LAN. Once enough TTs outside scene have been got beforehand, the rambling will become smoothly. When the distance that the viewpoint rambled through exceeds the size of TT, the whole terrain structure will be changed. As shown in Fig.4(b), when the viewpoint moved forward a TT size, the visible scene will become the

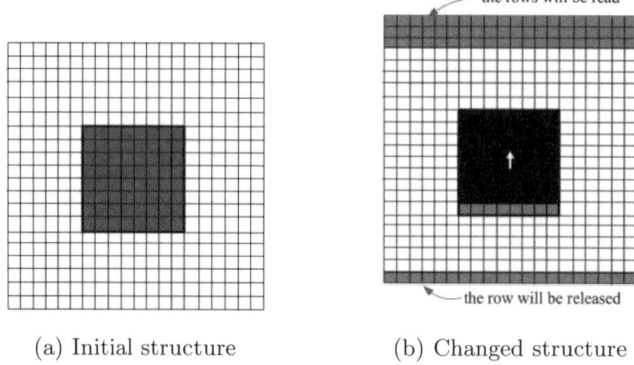

(a) Initial structure   (b) Changed structure

Fig. 4. Data structure of the terrain data

small black square area, at the same time , several rows of TTs on the top of the extent will be read, and the bottom row of TTs will be released from the RAM.

The second part of work tells us what the P2P agent system will do to search and transport the needed data. When firstly run into the scene, the 3DGIS application will search and transport the whole TTs of the terrain including the visible scene and the TTs outside the scene. When the 3DGIS application of the peer starts to run ,it will read the TTs indexes from the TTIF into RAM, at the same time the needed TTs indexes of the terrain are calculated, then application will search the needed TTs indexes in the index data read from the TTIF file, if there are all the indexes , it will read the DT and IT data form the DTDF file and the ITDF file ,if all the indexes are not found in the indexes from TTIF, it will make a request to the district server for the absent TTs. The district server peer will search and transport these TTs in the P2P network. The proposed algorithm will be described in section 6.

6   Algorithms for 3DGIS TTs Searching and Transportation

The speed of data searching and transportation is a most important factor to the on-line 3DGIS. If the needed data can not be obtained beforehand, the ramble speed of the 3DGIS will be slow down remarkably. However the mechanism based on traditional C/S can not fulfill the requirement of speed. Fortunately the P2P has the capability to solve the problem of the speed of data transportation. By constructing the high speed P2P network and agent with planning capability, we can realize the on-line 3DGIS with the ability of rambling smoothly. When the speed of data searching and transporting is taken into account, the Shortest-Latency Network (SLN) of P2P network must be constructed firstly[3], in the SLN agent's turnaround time between every two peers is the least, and then the partition of the peers of SLN is constructed, in every partition, the agent's turnaround time will not exceed a limit, in this way, the time of the data searching and transporting can be controlled, this is especially important for the on-line 3DGIS. Fig.5 shows the result SLN and partition of this paper.

In Fig.5(a), assuming the turnaround time δ of every partition is 200ms, the time of agent in every partition will never exceed 200ms. The TTs data searching and transportation algorithm is presented as follow:

1. Preprocessing: Construction of SLN: process the all-pairs except the DS peer shortest-path algorithm to construct an SLN network[3];
2. Partitioning: use δ to divide the district server peers into partitions. In fact, every server peer has its partitions that are different from the partitions of the other server peers. Fig.5(b) shows the partition of S_0.
3. When the simple peer $P_{0,1}$ need a list of TTs, $P_{0,1}$ will make a request to S_0 for these TTs .
4. S_0 will search the TTs indexes data, if there are needed TTs data on the simple peers in this district , the download progresses will be initiated to transport data from the simple peers via S_0 to $P_{0,1}$;

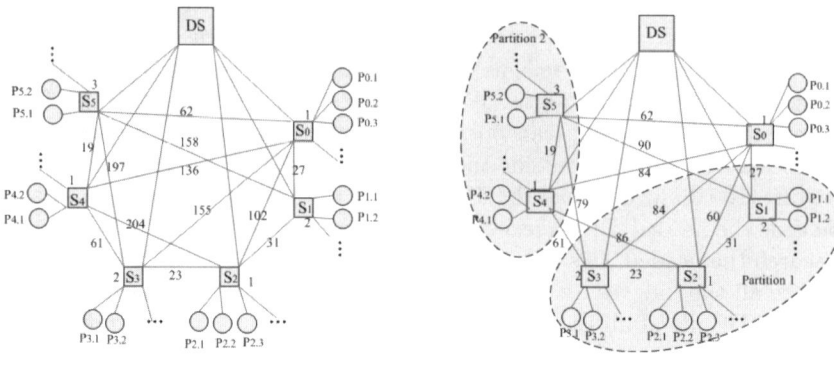

(a) Original network configuration   (b) SLN graph and partition

Fig. 5. SLN graph and partition of the P2P network

5. If there are some TTs not been found on S_0, S_0 will clone n agents and deploy them to every partition of S_0, n is equal to the amount of partitions. The district server peers of every partition will search the needed TTs data orderly. At last, the result will be returned to S_0 by mobile agent, and the algorithm will select the best path to download the data.
6. If there are TTs not been found yet after all the partitions have been searched, the 3DGIS application will download from the original data source server directly.

7   The 3DGIS Data Security in the P2P Agent System

The data security is an important content in the P2P agent system. If the security can not be ensured the data will be given away. The work of 3DGIS data security in this paper includes security of DEM data and image data .The encryption of DEM data is obtained by three algorithms: ①encryption of DT index; ②encryption of DEM elevation based on dynamic index related factors ;③encryption of data file. By the three algorithms, even though the data file has been cracked, the correct elevation of DT can not be obtained yet by reason of the rule of the index is stored on the original data source server. Encryption of DEM elevation is index related, the elevation of different DT is obtained by the product of the correct value and a different factor, the rule of index related factor is stored on the original data source server too. The encryption of IT data has another encryption algorithm of color index, called colormap, besides the encryption algorithm of DT data. The colormap is stored on the original data source server too. Consequently, the DT data has three layers of encryption algorithm, and the IT data has four layers of encryption algorithm. In this way, the security of 3DGIS data can be ensured.

8   Simulation

We have conducted the performance comparison between the P2P agent 3DGIS
and the 3DGIS based on C/S structure. We tested the two type of application
with different scene width in the wide area networks with different bandwidth.
Table.2 shows the minimum 3DGIS data transfer time of the two 3DGIS applications based on different bandwidth and specified scene width,and it shows
P2P agent 3DGIS performs better than the 3DGIS based on C/S structure.

Table 2. Minimum 3DGIS data transfer time of the two 3DGIS applications based on
different bandwidth and specified scene width

Wide Area Networks	the Minimum 3DGIS Data Transfer Time(sec) based on Specified Scene Width							
	C/S Structure based 3DGIS				P2P agent 3DGIS			
Bandwidth	8*8 TTs	10*10 TTs	14*14 TTs	16*16 TTs	8*8 TTs	10*10 TTs	14*14 TTs	16*16 TTs
56 Kbps	28.2	35.5	49.3	56.5	10.5	13.7	17.5	19.1
256 Kbps	5.1	7.4	10.2	11.8	2.4	3.0	3.7	4.2
1.54 Mbps	0.9	1.3	1.9	2.2	0.3	0.5	0.7	0.9
6.16 Mbps	0.2	0.25	0.36	0.45	0.1	0.15	0.26	0.34

When test the two applications with 8*8 TTs scene width in the wide area
networks with 256Kbps bandwidth, the data transfer time of the two applications
base on different amount of request users is shown in Fig.6 The test system is
shown in Fig.7.

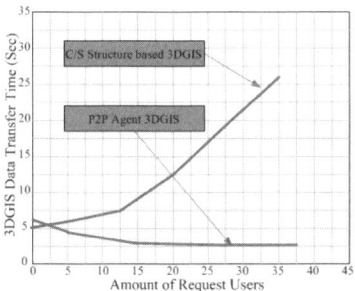
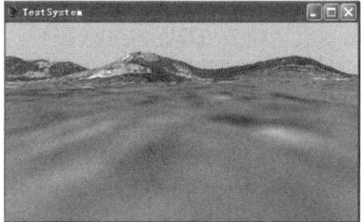

Fig. 6. The data transfer time of the two
applications base on different amount of request users

Fig. 7. Test system

9   Related Work

The study of on-line 3DGIS is popular now[7,8,9,10,11,12],most of these studies
are single-server based and constructed by the tools such as VRML and JAVA3D

etc, some of them obtained good performance too and can be used as references. There are also some 3D GIS system based on massive distribution of data over a large network ,for example Google Earth[13]. Compare with the approach proposed in this paper, Google Earth has a bigger width of terrain tile , it's suitable for browsing the terrain ,and the approach in this paper is able to construct complex terrain with many factors such as 3d-road,river,etc, and to roam smoothly in the scene. There are many researches of P2P network [4,5,6],and for realizing P2P service, the tools such as JXTA, SOCKET can be used .There are also some existing P2P service constructing API, for example FTKernel API. Many P2P systems have been developed, such as Napster, Gnutella and others. For the purpose of supporting mobile agent,many mobile agent environments have been developed,such as Aglets, D'agents, Mole. Several researches such as references[2,14,15,16,17,18] have studied the agent.

References

1. Takashi,Yamanoue. Sharing the Same Operation with a Large Number of Users Using P2P . Proc.IEEE ICITA2005 Sydney(2005)
2. Hui Tian ,Hong Shen. Mobile Agents Based Topology Discovery Algorithms and Modelling ,Proceedings of the 7th IEEE International Symposium on Parallel Architectures, Algorithms, and Networks (ISPAN'04).Hong Kong.(2004)214-219
3. Jin-Wook Baek , Heon Y. Yeom. An Approach to Provide Mobile Agents with Planning Capability for P2P Environment. Proceedings of the 2005 11th International Conference on Parallel and Distributed Systems (ICPADS'05). Fukuoka, Japan.1(2005):730-736
4. Takayuki HAMA ,Koichi ASATANI, Hidenori NAKAZATO. P2P Live streaming system with low signal interruption. Proceedings of the 18th International Conference on Advanced Information Networking and Application(AINA'04).Fukuoka, Japan(2004)
5. P. Reynolds , A. Vahdat.Efficient peer-to-peer keyword searching. Technical Report 2002, Duke University, CS Department, February 2002. Availableat: http://issg.cs.duke.edu/search/search.pdf.
6. Rodionov Maxim , Siu Cheung Hui. Intelligent Content-Based Retrieval for P2P Networks. Proceedings of the 2003 International Conference on Cyberworlds (CW'03).Singapore(2003)
7. Naphtali Rishe, Yanli Sun, Maxim Chekmasov, Andriy Selivonenko, Scott Graham . System Architecture for 3D TerraFly On-line GIS. Proceedings of the IEEE Sixth International Symposium on Multimedia Software Engineering (ISMSE'04).2004
8. Z.Chang ,Songnian Li.VRML-Based 3D Collaborative GIS:A Design Perspective. W2GIS 2004, Goyang, Korea, November 26-27,(2004)232-241
9. Chao Zhu, Eng Chong Tan, Tony Kai ,Yun Chan . 3D Terrain visualization for Web .http://www.gisdevelopment.net/technology/ip/ma03065.htm .july,2006
10. Ho-Geun Lee, Kyong-Ho Kim, and Kiwon Lee. Development of 3-Dimensional GIS Running on Internet, Geoscience and Remote Sensing Symposium Proceedings, IGARSS98 IEEE International 2(1998)1046-1049
11. Coors Volker. 3D-GIS in networking environments, Computers.Environment and Urban Systems 27(2003)345-357

12. Shu-Kai Yang, Ding-Zhou Duan, Ming-Fen Lin .Responsive Transmission of 3D
    Scenes over Internet .Third IEEE Pacific Rim Conference on Multimedia Hsinchu,
    Taiwan, December 16-18, **2532**(2002)1073-1079
13. Google Eearth Website http://earth.google.com .September 2006
14. Paulo Marques, Paulo Simes, Lus Silva, Fernando Boavida, Joo Silva . Provid-
    ing Applications with Mobile Agent Technology. Open Architectures and Network
    Programming Proceedings,IEEE,April, ,27-28,(2001)129-136
15. Ichiro Satoh .Network Processing of Mobile Agents, by Mobile Agents, for Mobile
    Agents .The 3rd International Workshop on Mobile Agents for Telecommunication
    Applications, Montreal, August, **2164**(2001)81-92
16. T.I.Wang .A Mobile Agent Carrier Environment for Mobile Information Re-
    trieval.11th International Conference on Database and Expert Systems Applica-
    tions,London, UK, September, **1873**(2000)634-643
17. H. Ku, G. W. R. Luderer, and B. Subbiah. An intelligent mobile agent frame-
    work for distributed network management.In Proc. IEEE GLOBECOM'97, (1997)
    160-164
18. V. A. Pham and A. Karmouch, Mobile software agents: An overview. IEEE Com-
    munications Magazine, (1998)26-37

Integrating Data from Maps on the World-Wide Web

Eliyahu Safra[1], Yaron Kanza[2,*], Yehoshua Sagiv[3,**], and Yerach Doytsher[1]

[1] Department of Transportation and Geo-Information, Technion, Haifa, Israel
{safra, doytsher}@technion.ac.il
[2] Department of Computer Science, University of Toronto, Toronto, Canada
yaron@cs.toronto.edu
[3] School of Engineering and Computer Science, The Hebrew University, Jerusalem, Israel
sagiv@cs.huji.ac.il

Abstract. A substantial amount of data about geographical entities is available on the World Wide Web, in the form of digital maps. This paper investigates the integration of such data. A three-step integration process is presented. First, geographical objects are retrieved from Maps on the Web. Secondly, pairs of objects that represent the same real-world entity, in different maps, are discovered and the information about them is combined. Finally, selected objects are presented to the user. The proposed process is efficient, accurate (*i.e.,* the discovery of corresponding objects has high recall and precision) and it can be applied to any pair of digital maps, without requiring the existence of specific attributes. For the step of discovering corresponding objects, three new algorithms are presented. These algorithms modify existing methods that use only the locations of geographical objects, so that information additional to locations will be utilized in the process. The three algorithms are compared using experiments on datasets with varying levels of completeness and accuracy. It is shown that when used correctly, additional information can improve the accuracy of location-based methods even when the data is not complete or not entirely accurate.

1 Introduction

Many maps are available on the World-Wide Web, providing information on geographical entities. The information consists of both spatial and non-spatial properties of the entities. Examples of spatial properties are location and shape of an entity. Examples of non-spatial properties are name and address. The goal of integrating two maps is to enable applications and users to easily access the properties that are available in either one of those maps. Another reason for integration is that some geographical entities may appear in only one of the maps. Integration increases the likelihood that for all the relevant entities, in a specified geographical area, objects that represent these entities are presented to the user.

An integration of two maps consists of the following three steps: extracting geographical objects from the maps, discovering pairs of objects that represent the same real-world entity in the two sources (such objects are called *corresponding objects*) and

---

* This author was supported by an NSERC grant.
** This author was supported by The Israel Science Foundation (Grant 893/05).

J.D. Carswell and T. Tezuka (Eds.): W2GIS 2006, LNCS 4295, pp. 180–191, 2006.

presenting the result to the user. This paper deals mainly with the second step of discovering corresponding objects. We use the term *matching algorithm* for an algorithm that discovers corresponding objects in two given datasets of geographical objects.

Methods for integrating data from the Web, and especially matching algorithms, should be able to cope with the following characteristics of the Web.

- Data on the Web is heterogeneous. This means that the same piece of information can have different forms in different sources. For example, in different sources, the name of a geographical entity can have different spellings or can be written in different languages. This makes it difficult for integration methods to use properties, such as names, for discovering corresponding objects. Another aspect of heterogeneity is incompleteness. Some attributes may not be available in some sources or not specified for some objects.
- Data may change frequently. For example, maps that contain hotels may also include reviews that are regularly added and updated by people who have stayed in those hotels. In such cases, the integration should be performed in real time, *i.e.*, when the user sends her request for information. Otherwise, the integrated data will not reflect the most recent changes in the sources. Consequently, an integration method for data on the Web must be efficient, especially if the method is used in a Web service that handles many requests concurrently.
- Data on the Web can be incorrect or inaccurate. Hence, on one hand, integration methods should rely mostly on object properties that are relatively accurate. On the other hand, this justifies using, in Web applications, approximation algorithms for matching, *i.e.*, *highly* (but not completely) *accurate* algorithms for discovering corresponding objects.

Because of the above reasons, in this paper we consider techniques that start with location-based matching algorithms and improve them. Relying primarily on locations has the following three advantages. First, locations are always available for spatial objects and their degree of accuracy can be determined relatively easily. Hence, location-based matching algorithms can be applied to objects from any pair of maps. Second, location-based methods are suitable for integration of heterogeneous data, since it is easy to compare a pair of locations even when they are stored or measured in different ways. Third, there exist efficient location-based matching algorithms.

Location-based matching algorithms that are both efficient and effective were presented in the past [2,3]. These algorithms use only locations for finding corresponding objects. Yet, in many cases, the accuracy of the integration can be improved significantly by using attributes of the integrated objects in addition to locations. This is especially important when dealing with data from the Web, where locations may be inaccurate. In this paper, we explain how to use properties of integrated objects to increase the effectiveness of location-based matching algorithms.

The main contributions of this paper are as follows. First, a complete process of integrating data from maps on the Web is presented. This process is efficient and general, in the sense that it can be applied to any pair of maps. Secondly, we show how, in addition to locations, attributes of the objects can be used in the integration process. Specifically, we present three new matching algorithms that use locations as well as additional information. Thirdly, we describe the results of thorough experiments, on datasets with different levels of accuracy and completeness, showing that additional information can improve the results of location-based matching algorithms, when that information is used appropriately.

The structure of the paper is as follows. In Section 2, we present our methods using a real-world example of integrating maps of hotels in the Soho area of Manhattan, New-York. We present our three new methods in Section 3. In Section 4, we provide the results of experiments that we conducted on both real-world data and syntactically generated data. Also, we compare our methods based on the experimental results. Finally, in Section 5, we discuses related work and conclude.

2   The Integration Process

We start by presenting our approach to integration of data from maps on the Web. We do that using an example that shows integration of information about hotels in the Soho area of Manhattan, New-York. The data sources that we used are Google Earth[1] and Yahoo Maps[2]. Google Earth is a service that provides a raster image of almost any part of earth. On top of the raster image, it shows information such as roads, hotels, and restaurants. In our example, we are interested in information about hotels. For hotels, Google Earth provides their names, which are shown as links that lead to additional information, *e.g.,* by following a link the address of the hotel is provided. A result of a search in Google Earth for hotels in Soho is depicted in Fig. 1.

Yahoo Maps provides road maps for some major cities in the world. As in Google Earth, maps include touristic information; however, in Yahoo, hotel names are not presented on the maps. Instead, a hotel is shown using an icon in the shape of a yellow square containing a red circle. The name of the hotel and additional information, such as the rank (*i.e.,* number of stars) and price are available for one hotel at a time, by clicking on the icon. Two possible reasons for not writing hotel names on the map are (*1*) making the presentation of the map simpler and easier to read (cartographic reasons), and (*2*) restricting the information released per each user request, so that applications will not be able to retrieve all the data from Yahoo to their local database (commercial reasons). A result of a search in Yahoo Maps for hotels in Soho is depicted in Fig. 2.

In the hotel scenario, it may seem a good solution to use a matching algorithm that considers as corresponding objects those pairs of hotels that have the same name. However, because names of hotels are not presented on maps from Yahoo, a matching based on names is problematic. Two other difficulties in using hotel names in a matching algorithm are the uncertainty in deciding whether two names refer to the same hotel and the presence of errors in the data. In our case, uncertainty is due to the existence of several hotels with similar names in the same vicinity. For instance, consider the following hotel names: "Grand Hotel," "Soho Grand Hotel" and "Tribeca Grand Hotel." Are these the names of three different hotels or of only two different hotels? Another case of uncertainty is when a hotel has more than one name. In the Soho area, the hotel named "Howard Johnson Express Inn," according to Google Earth, is named "Metro Three Hotel LLC" in Yahoo Maps, and indeed these are two names of the same hotel.

---

[1] http://earth.google.com
[2] http://maps.yahoo.com

Fig. 1. A Map from Google Earth   **Fig. 2.** A map from Yahoo Maps

In this work, we propose the following three-step integration process. *(1)* Retrieve the maps, extract relevant objects from the maps and compute the location of the objects. *(2)* Apply a matching algorithm for finding pairs of corresponding objects. *(3)* Display objects to the user (or return them as a dataset), where each pair of corresponding objects is represented by a single object. Objects that do not belong to any pair of corresponding objects may also be presented.

We now illustrate these steps using the Soho-hotels scenario. Initially, a search for hotels in Soho, New-York, was made in both Google Earth and Yahoo Maps, and the images of Fig. 1 and Fig. 2 were retrieved as a result. These two images were oriented using geo-referencing. Then, geographical objects were generated by digitizing the maps, that is, by identifying (in the raster images) icons of hotels and calculating their locations based on the geo-referencing. In this example, hotel names were inserted by a human user. In the future, we expect many maps on the Web to be in formats that computers can easily process without the need of human intervention. GML (Geographic Markup Language) [1] is an example of such a format.

The second step was to apply a matching algorithm to the two datasets that were extracted from the maps. The result of this step consists of pairs of objects that represent the same hotel, and of singletons representing hotels that appear in only one of the sources. More details about the matching algorithm will be given in the next section.

The final step of the integration is displaying to the user the pairs and singletons produced by the matching algorithm. Before providing the results, conditions can be used for selecting which objects to display. Note that filtering the results at this step makes it possible to apply conditions that use attributes from both sources.

3   Matching Algorithms

The most involved part of an integration process is the discovery of corresponding objects, *i.e.,* the matching algorithm. Several matching algorithms that use only the location of objects were proposed in the past [2,3]. We now present three new algorithms that are built upon existing location-based algorithms and use attributes of objects for improving the matching.

3.1 Framework

First, we present our framework. A *dataset* is a collection of geographical objects that are extracted from a given map. Each object represents a single real-world *geographical entity* and has a point location. (For an object that has a polygonal shape, we consider the center of mass of the polygonal shape to be the point location of the object.) The distance between two objects is the Euclidean distance between their point locations. We denote by $distance(a, b)$ the distance between two objects a and b.

An object may have, in addition to location, attributes that contain information about the entity that the object represents. We distinguish between two types of attributes. An attribute I of objects in a dataset A is *unique* if every two objects in A have different values for I, *i.e.*, I is a candidate key. We consider I as *non-unique* if there can be two objects in A that have the same value for I. For example, in a dataset of hotels, the name of a hotel is a unique attribute, since it is unlikely that two hotels in the same vicinity will have the same name. We consider rating (number of stars) as non-unique, because two proximate hotels may have the same number of stars. When locations of objects are not accurate, we can improve a basic matching algorithm by using additional attributes. If the additional information is correct, a unique attribute can be used for discovering pairs of corresponding objects that the basic algorithm fails to match. Both unique and non-unique attributes can be used for detecting pairs of non-corresponding objects that are, wrongly, deemed corresponding by a matching algorithm.

In integration of maps, locations of objects are not accurate, because the process of extracting objects and computing their locations, by digitizing an image, introduces errors. Furthermore, maps on the Web may not be accurate to begin with. Thus, given two datasets A and B that are extracted from two maps, two corresponding objects $a \in A$ and $b \in B$ may not have the same location. Yet, for each dataset, errors are normally distributed with some standard deviation σ. So, for 98.8% of the objects, their distance from the real-world entity that they represent is less than or equal to 2.5σ. Hence, for 98.8% of the pairs $\{a, b\}$ of corresponding objects, it holds that $distance(a, b) \leq \beta$, where $\beta = 2.5\sqrt{\sigma_A^2 + \sigma_B^2}$ is the *distance bound* of A and B (σ_A and σ_B are the standard deviations of the error distributions in A and B, respectively). In our algorithms, pairs $\{a, b\}$ with $distance(a, b) > \beta$ are never deemed corresponding objects.

A matching algorithm receives a pair of datasets A and B and returns two sets P and S. The set P consists of pairs $\{a, b\}$, such that $a \in A$ and $b \in B$ are likely to be corresponding objects. The set S consists of singletons $\{s\}$ (where $s \in A \cup B$) such that, with high likelihood, s does not have a corresponding object. Location-based matching algorithms compute the sets P and S according to the distance between objects.

3.2 The New Matching Algorithms

We now describe three new algorithms that receive an existing matching algorithm $\mathcal{M}$ and improve it by using the information provided by some specified attributes. We divide the input to these algorithm into two parts. One part consists of two datasets A and B that should be joined. The second part consists of $\mathcal{M}$, a set X of the given attributes and, for the third algorithm, an additional factor ϕ. We denote by P and S the set of pairs and the set of singletons, respectively, that the algorithms return. The pseudocode of all three algorithms is presented in Fig. 3.

$Pre\text{-}D_{[\mathcal{M},X]}(A,B)$

Parameters: A matching algorithm $\mathcal{M}$, a set of *unique* attributes X
Input: Datasets A and B
Output: A set P of pairs and a set S of singletons
1: $P \leftarrow \emptyset, S \leftarrow \emptyset, A' \leftarrow A, B' \leftarrow B$
2: let β be the distance bound of A and B
3: **for each** $a \in A$ and $b \in B$ such that $a.x = b.x$ for some attribute $x \in X$ **do**
4:    **if** $distance(a,b) \leq \beta$ **then**
5:        $P \leftarrow P \cup \{a,b\}$
6:        $A' \leftarrow A' - \{a\}, B' \leftarrow B' - \{b\}$
7: $(P',S') \leftarrow \mathcal{M}(A',B')$
8: $P \leftarrow P \cup P', S \leftarrow S'$
9: **return** (P,S)

$Post\text{-}R_{[\mathcal{M},X]}(A,B)$

Parameters: A matching algorithm $\mathcal{M}$, a set of attributes X
Input: Datasets A and B
Output: A set P of pairs and a set S of singletons
1: $(P,S) \leftarrow \mathcal{M}(A,B)$
2: **for each** $\{a,b\} \in P$ such that $a.x \neq b.x$ for some attribute $x \in X$ **do**
3:    $P \leftarrow P - \{a,b\}$
4: **return** (P,S)

$Pre\text{-}F_{[\mathcal{M},X,\phi]}(A,B)$

Parameters: A matching algorithm $\mathcal{M}$, a set of *non-unique* attributes X, a factor ϕ
Input: Datasets A and B
Output: A set P of pairs and a set S of singletons
1: $P \leftarrow \emptyset, S \leftarrow \emptyset$
2: let $distance_n(x,y)$ be a new distance function that, initially, is equal to $distance(x,y)$
3: **for each** $a \in A$ and $b \in B$ such that $a.x \neq b.x$ for some attribute $x \in X$ **do**
4:    $distance_n(a,b) \leftarrow \phi \cdot distance(a,b)$
5: let $\mathcal{M}_n$ be the matching algorithm $\mathcal{M}$ when run using the distance function $distance_n(x,y)$ instead of using the Euclidean distance function $distance(x,y)$
6: $(P,S) \leftarrow \mathcal{M}_n(A,B)$
7: **return** (P,S)

Fig. 3. The algorithms Pre-process detection, Post-process removal and Pre-process factorizing

Pre-process detection (*Pre-D*)

The *Pre-D* algorithm uses unique attributes for detecting corresponding objects, and then it calls another matching algorithm on the remaining objects. The algorithm has two steps.

1. For each pair of objects $a \in A$ and $b \in B$, such that a and b have the same value for some unique attribute of X and the distance between them does not exceed the

distance bound of A and B, the pair $\{a, b\}$ is added to P, a is removed from A and b is removed from B.

2. The matching algorithm $\mathcal{M}$ is applied to the remaining objects of A and B. Upon termination, the pairs of the result are added to P and the singletons—to S.

Post-process removal (*Post-R*)

The *Post-R* algorithm uses a set of attributes X for detecting pairs of objects that are erroneously matched by another algorithm. The *Post-R* algorithm has two steps.

1. The matching algorithm $\mathcal{M}$ is applied to A and B. The result is a set P of pairs and a set S of singletons.
2. For each pair of objects $\{a, b\}$ in P, such that a and b have different values for some attribute of X, the pair $\{a, b\}$ is removed from P.

Pre-process distance factorization (*Pre-F*)

The *Pre-F* algorithm uses a set X of non-unique attributes as follows. For every pair of objects $a \in A$ and $b \in B$ that have different values for some attribute of X, the distance between a and b is multiplied by the given factor $\phi > 1$. Note that increasing the distance between objects lowers the probability that they will be matched by a location-based algorithm. The algorithm $\mathcal{M}$ uses the new distances to join A and B.

In our experiments, we tested eight different combinations of the above algorithms. Suppose that the set Y contains the shared attributes of two datasets A and B. Let $unique(Y)$ and $non\text{-}unique(Y)$ be the sets of unique and non-unique attributes of Y, respectively. Given a location-based matching algorithm $\mathcal{M}$, the following are the eight possible ways of computing the matching of A and B.

1. Use only the location based algorithm $\mathcal{M}$, *i.e.*, return $\mathcal{M}(A, B)$.
2. Use *Post-R* with $\mathcal{M}$. That is, return $Post\text{-}R_{[\mathcal{M}, Y]}(A, B)$.
3. Use *Pre-D* with $\mathcal{M}$. That is, return $Pre\text{-}D_{[\mathcal{M}, unique(Y)]}(A, B)$.
4. Combine *Pre-D* and *Post-R*, *i.e.*, return $Post\text{-}R_{[Pre\text{-}D_{[\mathcal{M}, unique(Y)]}, Y]}(A, B)$.
5. Use *Pre-F* with $\mathcal{M}$. That is, return $Pre\text{-}F_{[\mathcal{M}, non\text{-}unique(Y), \phi]}(A, B)$.
6. Combine *Post-R* with *Pre-F*, *i.e.*, return $Post\text{-}R_{[Pre\text{-}F_{[\mathcal{M}, non\text{-}unique(Y), \phi]}, Y]}(A, B)$.
7. Combine *Pre-D* with *Pre-F*. That is, return the result of the following expression:
   $Pre\text{-}D_{[Pre\text{-}F_{[\mathcal{M}, non\text{-}unique(Y), \phi]}, unique(Y)]}(A, B)$.
8. Combine all the three methods by applying *Pre-F*, *Pre-D*, $\mathcal{M}$ and, finally, *Post-R*, *i.e.*, return $Post\text{-}R_{[Pre\text{-}D_{[Pre\text{-}F_{[\mathcal{M}, non\text{-}unique(Y), \phi]}, unique(Y)]}, Y]}(A, B)$.

3.3 Computing the Distance Bound

Applying a matching algorithm requires knowing the distance bound β (or an approximation of it). The approximation of β is computed based on approximations of σ_A and σ_B—the standard deviations of the error distributions in the integrated datasets (see Section 3.1). The values σ_A and σ_B (we also call them the *errors* of the datasets) are sometimes provided with the maps, and in other cases we need to estimate them.

The error of a dataset is caused by errors in the procedure of collecting and processing the geographical data. The procedure is different when generating raster (imagery) maps and when vector (feature based) maps are produced. (See [11] for more detailed descriptions of these procedures.)

Raster maps are typically generated from satellite or aerial photographs. There are three main causes of error in the process of creating raster maps. First, errors are introduced when the photos are orthorectified *i.e.,* when correcting the photos to accurately represent the surface of the earth. Second, the size of the pixels in the photo affects the error. Currently, a resolution of 70cm per pixel at nadir is common in satellite imagery (*e.g.,* in the two main high-resolution commercial earth-observation satellites IKonos and QuickBird). The first two factors are relatively small and the main cause of error is the third factor which is the accuracy of the geo-referencing process *i.e.,* the accuracy of matching earth coordinates to the imagery. The accuracy of the geo-referencing depends on the existence and accuracy of reference points. When no reference points exist, the accuracy is about 10 meters, while when there are reference points, the accuracy is about 1–10 meters, depends on the accuracy of the reference points. Extracting features from the raster image (*e.g.,* identifying the location of an hotel) also introduces an error which is approximately the number of pixels of the error in the extraction process multiplied by the resolution.

Vector maps are usually created either by governmental mapping agencies, or by commercial companies, according to agreed mapping standards. The standards define accuracy requirements that depend on the map scale. Typically, for urban areas, map scales are between 1/1000–1/10000. Normally, the required accuracy for such scales is about 0.3–0.4mm. For example at a scale of 1/5000, the error is about 1.5–2 meters.

3.4   Measuring the Quality of the Result

We use *recall* and *precision* to measure the accuracy of a matching algorithm. Remember that the result of a matching algorithm consists of sets (singletons and pairs). A set is correct if it is either a pair of corresponding objects or a single object that has no corresponding object. Given the result of a matching algorithm, the recall is the ratio of the number of correct sets in the result to the number of all correct sets. For example, a recall of 0.8 means that 80% of the correct sets appear in the result. The precision is the ratio of the number of correct sets in the result to the number of sets in the result. For example, a precision of 0.9 means that 90% of the sets in the result are correct.

In our experiments, we knew exactly which sets were correct and, hence, were able to determine the precision and recall. For synthetic data, all the information about the data was available to us. For real-world data, we determined the correct sets manually, using all the available information.

4   Experiments

In this section, we describe the results of extensive experiments on both real-world and synthetically generated data. The goal of our experiments was to compare the eight combinations, presented in Section 3.2, over data with varying levels of inaccuracy and incompleteness. We also wanted to determine by how much our methods improve existing location-based algorithms. For that, we tested the effect of our methods on the following three location-based algorithms: nearest-neighbor (NN), mutually-nearest (MUTU) and normalized-weights (NW); see [3] for a description of these algorithms.

4.1   Tests on Real-World Data

We present the results of integrating the maps of hotels in Soho as described in Section 2. The Google-Earth map presents 28 hotels and the map from Yahoo Maps presents 39 hotels and inns. A total number of 44 hotels and inns appear in these sources, where 21 hotels appear in both of the sources while 23 appear in only one source. For both sources, we used an error σ of 100 meters because identifying the location of an hotel based on an icon is highly inaccurate.

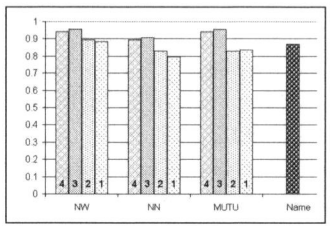

Fig. 4. Tests on real-world data

Figure 4 shows the harmonic mean of the recall and precision (HRP) for the three location-based algorithms (NW, MUTU, NN). Each one of the three algorithms was tested according to the first four combinations of Section 3.2. (The other four combinations are not applicable, since the only attribute, hotel name, is unique.) The third combination, *Pre-D*, is clearly the best for each of the three algorithms. It is slightly better than the fourth combination, which includes both *Pre-D* and *Post-R*, since the attribute hotel name is not always accurate (e.g., one hotel has different names in the two sources). For comparison, Figure 4 also shows the result of matching just according to hotel names. Note that for combinations 2–4, the process was semi-automatic, since hotel names do not appear in Yahoo Maps.

4.2   Tests on Synthetic Data

In order to test our methods on data with varying levels of accuracy and incompleteness, we randomly generated synthetic datasets using a two-step process. First, the real-world entities are generated. The locations of these entities are randomly chosen, according to a uniform distribution, in a square area. Each entity has one unique attribute U and one non-unique attribute N with randomly-chosen values. The non-unique attribute has five possible values (as for the number of stars of a hotel). In the second step, the objects in each dataset are generated. Each object is associated with a distinct entity and its location is chosen with an error that is normally distributed (relative to the location of the entity). In each dataset, different objects correspond to distinct entities. For each object, the attribute U has either the same value as in the corresponding entity, null (for incompleteness) or an arbitrary random value (for inaccuracy). We denote by $c(U)$ the percentage of objects that have a non-null value for U and by $a(U)$ the percentage of objects that have either the correct value or null. Values are similarly assigned to N.

We present the results of two tests. In Test I, the values of the attributes are either accurate or missing (i.e., null). In Test II, all the objects have values for U and N, but some of those values are inaccurate. In both tests, there are 1000 entities in a square

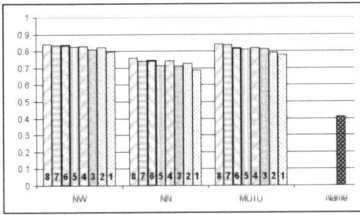

Fig. 5. Results of Test I

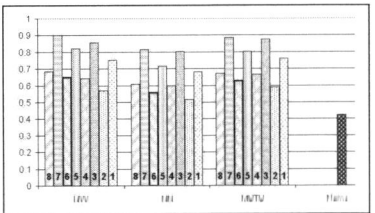

Fig. 6. Results of Test II

area of 1350×1350 meters with a minimal distance of 15 meters between entities. Each dataset has 750 objects that are randomly chosen for 750 entities using a standard deviation of $\sigma = 12$ meters for the error distribution. In Test I, the attributes in each dataset have either the correct values or nulls as follows: $a(U) = a(N) = 100\%$, $c(U) = 40\%$ and $c(N) = 60\%$. That is, only 40% of the objects have the correct value for the unique attribute and only 60% of the objects have the correct value for the non-unique attribute (if the value is not the correct one, then it is null). In Test II, attributes always have non-null values but not necessarily the correct ones, i.e., $c(U) = c(N) = 100\%$ and $a(U) = a(N) = 80\%$.

In Test I and Test II, we tried the eight combinations of Section 3.2 with each of the three algorithms. The results, depicted in Fig. 5 and. 6, show the harmonic mean of the recall and precision for the eight combinations involving each algorithm. Each bar is for the combination identified by the number on that bar. For comparison, we also show the result obtained by a matching algorithm that only uses the unique attribute (Name).

Test I shows that when information is partial but accurate, the eighth combination that uses all of the three algorithms (*Pre-D*, *Post-R* and *Pre-F*) is the best. Test II shows that when information is inaccurate, *Post-R* is not effective (as was also the case for the real-world data) and it is better to use just *Pre-D* and *Pre-F* (the seventh combination).

Figures 7 and 8 show the performance of the NW method for varying levels of completeness and accuracy. In Figure 7, the accuracy varies, *i.e.*, $a(U) = a(N) = 70\% \ldots 100\%$, and the completeness is fixed, *i.e.*, $c(U) = c(N) = 100\%$. In Figure 8, the completeness varies, *i.e.*, $c(U) = c(N) = 40\% \ldots 100\%$, and the accuracy is fixed, *i.e.*, $a(U) = a(N) = 100\%$. In each graph, the serial number refers to the combination that produced the graph. Note that the results of only 6 methods (1,2,3,5,7,8) are presented, since the other two are inferior.

The followings are our conclusions from the tests.

1. When there is a *unique* attribute, it is always good to identify pairs and remove them from the matching algorithm (Method 2).
2. When there is a non-unique attribute, it is always good to use factorized distance (Method 5).
3. Although additional information improves the quality of the results, the main factor that determines the quality is still the location-based algorithm.
4. When the attributes are not accurate, using the additional information before the matching improves the quality of the result. But using it after the location-based matching has a negative effect, for the following reason. While there is only a low

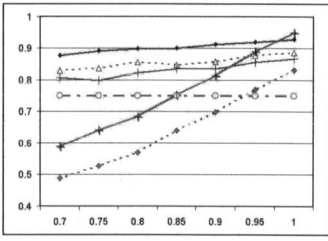   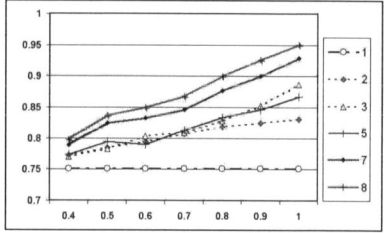

Fig. 7. Results of NW for varying accuracy   **Fig. 8.** Results of NW for varying completeness

probability that two proximate yet non-corresponding objects have the same value for a unique attribute,there is a considerably higher probability that two corresponding objects have different values for some unique attribute.

The tests show that in all cases using additional attribute before applying a location-based matching algorithm improves the quality of the results. Applying additional information at the end yields an improvement only if that information is accurate.

5   Conclusion

Traditionally, integration of geo-spatial data is being done using map conflation [13,6]. However, map conflation is not efficient since whole maps are integrated, not just selected objects. Thus, conflation is not suitable for Web applications or in the context of mediators [4,12,19,20] where users request answers to specific queries. Integrating spatial datasets using only geometrical or topological properties [2,3,14] or using only alpha numeric attributes [9,10], both do not use all the available information but can be combined using the approach we introduced in this paper.

Other approaches use both spatial and non-spatial attributes (*e.g.* [7,15,17]). However, these approaches occasionally remain on the schema level, rather than actually matching the objects, such as [7], or has large computation time as [15,17].

In this work we showed how data from maps on the Web can be integrated using location-based algorithms, and how to utilize information additional to location when such information exists. We presented three new matching algorithms and tested them on data with varying levels of incompleteness and inaccuracy. Interestingly, our experiments show that when the additional information is accurate it should be used both before and after the location-based matching process. When the additional information is not very accurate, the information should be used only prior to the location-based matching process. Our experiments show that the new algorithms improve the existing location-based matching algorithms.

References

1. Geographic Markup Language (GML). http://www.opengeospatial.org/standards/gml.
2. C. Beeri, Y. Doytsher, Y. Kanza, E. Safra, and Y. Sagiv. Finding corresponding objects when integrating several geo-spatial datasets. In *ACM-GIS*, pages 87–96, 2005.

3. C. Beeri, Y. Kanza, E. Safra, and Y. Sagiv. Object fusion in geographic information systems. In *VLDB*, pages 816–827, 2004.
4. O. Boucelma, M. Essid, and Z. Lacroix. A WFS-based mediation system for GIS interoperability. In *ACM-GIS*, pages 23–28, 2002.
5. T. Bruns and M. Egenhofer. Similarity of spatial scenes. In *SDH*, pages 31–42, Delft (Netherlands), 1996.
6. M. A. Cobb, M. J. Chung, H. Foley, F. E. Petry, and K. B. Show. A rule-based approach for conflation of attribute vector data. *GeoInformatica*, 2(1):7–33, 1998.
7. T. Devogele, C. Parent, and S. Spaccapietra. On spatial database integration. In *IJGIS, Special Issue on System Integration*, 1998.
8. F. T. Fonseca and M. J. Egenhofer. Ontology-driven geographic information systems. In *ACM-GIS*, pages 14–19, Kansas City (Missouri, US), 1999.
9. L. Gravano, P. G. Ipeirotis, H. V. Jagadish, N. Koudas, S. Muthukrishnan, and D. Srivastava. Approximate string joins in a database (almost) for free. In *VLDB*, pages 491–500, 2001.
10. L. Gravano, P. G. Ipeirotis, N. Koudas, and D. Srivastava. Text joins in an RDBMS for web data integration. In *Proceedings of the 12th international conference on World Wide Web*, pages 90–101, 2003.
11. J. C. McGlone. *Manual of Photogrammetry, Fifth Edition*. American Society of Photogrammetry and Remote Sensing, 2004.
12. Y. Papakonstantinou, S. Abiteboul, and H. Garcia-Molina. Object fusion in mediator systems. In *VLDB*, pages 413–424, 1996.
13. A. Saalfeld. Conflation-automated map compilation. *IJGIS*, 2(3):217–228, 1988.
14. A. Samal, S. Seth, and K. Cueto. A feature based approach to conflation of geospatial sources. *IJGIS*, 18(00):1–31, 2004.
15. M. Sester, K. H. Anders, and V. Walter. Linking objects of different spatial data sets by integration and aggregation. *GeoInformatica*, 2(4):335–358, 1998.
16. H. Uitermark, P. Van Oosterom, N. Mars, and M. Molenaar. Ontology-based geographic data set integration. In *Proceedings of Workshop on Spatio-Temporal Database Management*, pages 60–79, Edinburgh (Scotland), 1999.
17. V. Walter and D. Fritsch. Matching spatial data sets: a statistical approach. *IJGIS*, 13(5):445–473, 1999.
18. J. M. Ware and C. B. Jones. Matching and aligning features in overlayed coverages. In *ACM-GIS*, pages 28–33, 1998.
19. G. Wiederhold. Mediators in the architecture of future information systems. *Computer*, 25(3):38–49, 1992.
20. G. Wiederhold. Mediation to deal with heterogeneous data sources. In *Introperating Geographic Information Systems*, pages 1–16, 1999.

Web-Based Cluster Analysis for the Time-Series Signature of Local Spatial Association

Jae-Seong Ahn[1], Yang-Won Lee[2,*], and Key-Ho Park[1]

[1] Department of Geography, College of Social Sciences, Seoul National University, Korea
[2] Center for Spatial Information Science, The University of Tokyo, Cw-503 IIS Bldg., 4-6-1
Komaba, Meguro-ku, Tokyo 153-8505, Japan
jwlee@iis.u-tokyo.ac.jp

Abstract. We propose a method for modeling the time-series of local spatial association in geographical phenomena and implement a Web-based statistical GIS for the time-series analysis using client-provided dataset. In order to examine the pattern of time-series and classify similar ones on a cluster basis, we employ Moran scatterplot and extend it to time-series Moran scatterplot accumulated over a certain span of time. Using the time-series Moran scatterplot, we develop similarity measures of "state sequence" and "clustering transition" for the time-series of local spatial association. If we connect n corresponding points of a region on the time-series Moran scatterplot, the connected line composed of n nodes and n-1 edges forms a time-series signature of local spatial association for the region. From the similarity matrix of the time-series signatures, we generate a map of the clustered classification of changing regions. These analytical functionalities of cluster analysis on the time-series of local spatial association are implemented in a Web-based GIS using XML Web Services.

Keywords: Web-based Statistical GIS, Local Spatial Association, Cluster Analysis.

1 Introduction

The concept of spatial association may represent one of the core philosophies in geography [27, 18]: "Everything is related to everything else, but near things are more related than distant things." The relationship or association among geographic entities has been modeled in statistical measures at global or local scale. While the global measures of spatial association [19, 9, 10] present the overall degree of spatial dependence among regions in the whole area, the local measures of spatial association [2, 20] provide the information about the individual degree of spatial dependence of a region and its neighbors. The local spatial association can be extended to cluster formation, in that similar regions have more possibility to be located adjacent to each other. Besides, since the relationship of a region and its neighbors changes by time, the changes in local spatial association need to be approached on a time-series basis. In temporal context, a cluster analysis based on the similarity of the time-series of

---

* Corresponding author.

J.D. Carswell and T. Tezuka (Eds.): W2GIS 2006, LNCS 4295, pp. 192–201, 2006.
© Springer-Verlag Berlin Heidelberg 2006

local spatial association could facilitate the capturing of common changes in spatially associated regions.

One of the recent trends in computing environment of spatial data analysis may be Web-based GIS for open access to analytical functionalities. Beyond the previous emphasis of the Web-based GIS on map delivery, cartographic presentation, and providing of geographical information [14, 22, 13, 15, 28], more specialized spatial analytical capabilities are implemented on the Web nowadays [4, 6, 16] including exploratory spatial data analysis [1, 21, 25, 26, 4] and spatial modeling [11, 24, 29]. In addition to these contributions, the analytical functionalities allowing for client-provided dataset can be another contribution. So far, Web-based analytical GIS have mainly dealt with server-provided dataset, but the incorporation of client-provided dataset may elevate the interactiveness and extensibility that ensure more open access to spatial data analysis on the Web.

The objective of this paper is to propose a method for modeling the time-series of local spatial association in geographical phenomena and implement a Web-based statistical GIS for the time-series analysis using client-provided dataset. In order to examine the pattern of time-series and classify similar ones on a cluster basis, we employ Moran scatterplot and extend it to time-series Moran scatterplot accumulated over a certain span of time. Using the time-series Moran scatterplot, we develop similarity measures of "state sequence" and "clustering transition" for the time-series of local spatial association. If we connect n corresponding points of a region on the time-series Moran scatterplot, the connected line composed of n nodes and n-1 edges forms a time-series signature of local spatial association for the region. From the similarity matrix of the time-series signatures, we generate a map of the clustered classification of changing regions. These analytical functionalities of cluster analysis on the time-series of local spatial association are implemented in a Web-based GIS using XML Web Services.

While the background and objective of this paper being introduced in this section, we examine related work to spatial association in section 2. As to the components of the proposed method, section 3 describes the time-series signature of local spatial association, similarity measures for the time-series signature, cluster analysis for the time-series signature, and Web-based GIS implementation for these analytical functionalities. Section 4 demonstrates the implemented system by use of a client-provided dataset: elderly population ratio of 65 administrative units in Seoul Metropolitan Area, 1995-2004. Section 5 concludes the paper with a summary and implications of our work.

2   Spatial Association

Global measures of spatial association (or autocorrelation) such as Moran's I [19], Geary's C [9], and G statistics of Getis & Ord [10] derive a single value representing a study area. However, since the global value may not be universally applicable throughout the whole area, local measures of spatial association alternatively examine the spatial dependence among subset regions focusing on the variation within the study area [7, 8, 5]. As local indicators of spatial association (LISA), local Moran's I, local Geary's C [2], and local G statistics of Ord & Getis [20] provide the local values for each subset region.

Local Moran's *I*, the most commonly used local indicator is extended to Moran scatterplot [3] that assesses the local instability in spatial association in a study area. As in Figure 1, the horizontal axis denotes Z-score of each subset region, and the vertical axis denotes spatially lagged Z-score, namely, mean of the neighbors' Z-score of the corresponding region. The four quadrants of Moran scatterplot are interpreted as in Table 1. The Moran Scatterplot allows for a visualization of several geographical aspects of the distribution at a given point in time. First, as the spatial-lag pairs become concentrated in quadrant I/III or quadrant II/IV, the overall level of spatial association in the distribution strengthens. Secondly, outliers in terms of spatial-lag pairs that deviate from the overall trend can be easily identified. Thirdly, clusters of local spatial association can also be identified in the scatterplot [23].

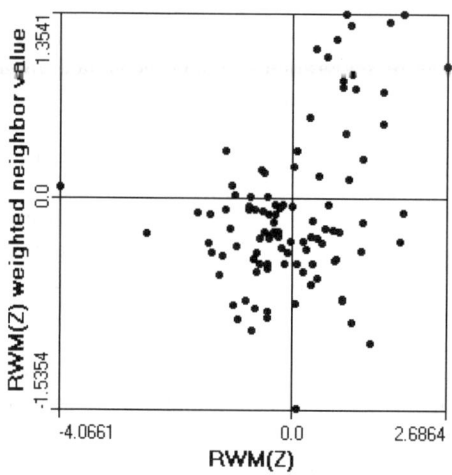

Fig. 1. Moran scatterplot for assessing the local pattern of spatial association

Table 1. Interpretation of Moran scatterplot

State	Quadrant	Autocorrelation	Interpretation
HH	I	Positive	Cluster - "I'm high and my neighbors are high."
LH	II	Negative	Outlier - "I'm a low outlier among high neighbors."
LL	III	Positive	Cluster - "I'm low and my neighbors are low."
HL	IV	Negative	Outlier - "I'm a high outlier among low neighbors."

If considering temporal change of the state of Moran scatterplot, a state sequence {LL → LL → LH → LH → HH → HH}, for instance, implies that "I was low like my neighbors (LL). While I was still low, my neighbors became high (LH). Then, I became high like my neighbors (HH)." In order to examine such changes, Rey [23] proposes a space-time transition measure using Moran scatterplot and Markov chain. This method provides a refined analysis on the probability of the regional change in relation to the change of neighbors. Five transition types are defined in the analysis. Type 0 denotes no move of both a region and its neighbors. Type I denotes a relative move of only the region, such as HH → LH, HL → LL, LH → HH, and LL → HL.

Type II denotes a relative move of only the neighbors, such as HH → HL, HL → HH, LH → LL, and LL → LH. Type III denotes a move of both a region and its neighbors to a different side: Type IIIA includes joint upward move (LL → HH) and joint downward move (LL → HH) whereas Type IIIB includes diagonal switch such as HL → LH and LH → HL. Since these transition types are derived on a time-span basis and incorporated in a discontinuous Markov chain, each transition is taken as an individual observation. For example, t_0 → t_1 → t_2 → ... → t_7 → t_8 → t_9 includes nine individual observations for the time-span {t → t+1}, and each observation participates in calculating the probability of regional change by transition type, regardless of time order.

3 Proposed Method

Since the above space-time transition measure is based on a discontinuous Markov chain, the probability of regional change by transition type may not sufficiently reflect the sequential aspect of regional change. As an alternative to this, we deal with the regional change using a time-ordered sequence of region's state that continuously connects the state of local spatial association at certain intervals. More importantly, our method provides a clustered classification of changing regions according to the similarity of the time-series signature of local spatial association. These analytical functionalities of cluster analysis on the time-series of local spatial association are implemented using XML Web Services that can ensure standardized open access to Web-based statistical GIS.

3.1 Time-Series Signature of Local Spatial Association

In order to examine the pattern of the time-series of local spatial association, we employ Moran scatterplot and extend it to the time-series Moran scatterplot that is accumulated over a certain span of time. If we connect n corresponding points of a region on the time-series Moran scatterplot, the connected line composed of n nodes and n-1 edges forms a time-series signature of local spatial association for the region. While the node of a region's signature represents the state of local spatial association at a given point in time, the edge of a region's signature represents the transition of local spatial association in a given period of time. For the notations of a node's state of local spatial association, we use I/II/III/IV after four quadrants of Moran scatter-plot. In addition, an edge connecting two nodes has the information about "clustering transition," such as upward/downward clustering and declustering. We set up six notations for the "clustering transition" as in Table 2.

Upward clustering (UC) includes the transitions such as LH → HH and HL → HH. Downward clustering (DC) includes the transitions such as LH → LL and HL → LL. Upward declustering (UD) includes the transitions that a region (relatively) moves up while its neighbors (relatively) move down, such as HH → HL and LL → HL. Downward declustering (DD) includes the transitions that a region (relatively) moves down while its neighbors (relatively) move up, such as HH → LH and LL → LH. In addition, joint upward clustering (JU) denotes LL → HH, and joint downward clustering (JD) denotes LL → HH transition. No change in the transition is represented as (O).

Table 2. Notations for "clustering transition" of local spatial association

	I (HH$_{i+1}$)	II (LH$_{i+1}$)	III (LL$_{i+1}$)	IV (HL$_{i+1}$)
I (HH$_i$)	O	**DD**	**JD**	**UD**
II (LH$_i$)	**UC**	O	**DC**	**UD**
III (LL$_i$)	**JU**	**DD**	O	**UD**
IV (HL$_i$)	**UC**	**DD**	**DC**	O

3.2 Similarity Measures for Time-Series Signature of Local Spatial Association

Every region has its own time-series signature of local spatial association. Some of them may be similar or may be not. Similarity measure for the time-series signature is a key to the cluster analysis of local spatial association because the similarity explains as to the cluster formation of the regions that have experienced similar history in terms of local spatial association. In order to obtain the similarity matrix of region pairs from the time-series signature of each region, we employ Levenshtein metric, the most commonly used method for sequence comparison of categorical data [12]. The Levenshtein metric is defined as the minimum number of edit operation (insertion/deletion/substitution) needed to transform one sequence into the other, namely, a unit cost for the edit operations [17]. Figure 2(a) is a pseudocode for the function *LevenshteinDistance* that takes two sequences, *seq1* of length *lenSeq1*, and *seq2* of length *lenSeq2*, and computes the Levenshtein metric between them. Figure 2(b) is a calculation example of the Levenshtein metric between {III-III-II-I-I-I} and {III-II-II-I-I-IV}.

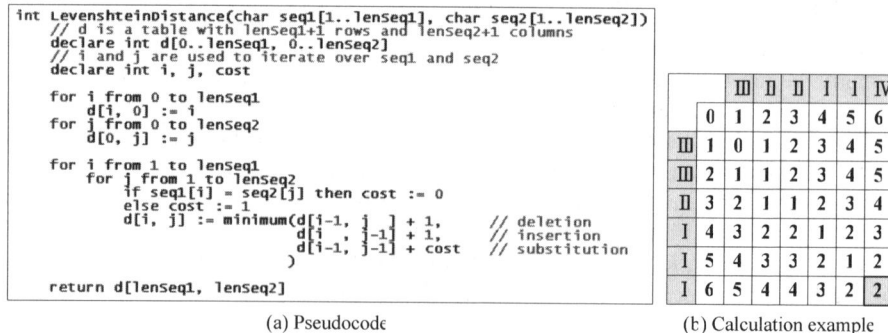

(a) Pseudocode   (b) Calculation example

Fig. 2. Levenshtein metric for the sequence comparison of categorical data

Using the Levenshtein metric, the similarity measures for the time-series signature of local spatial association indicates the dissimilarity of each region pair in terms of "state sequence" (using the notations I/II/III/IV) and "clustering transition" (using the notations UC/DC/UD/DD/JU/JD). The black point in Figure 3(a) represents the state at a given point in time, and the thick line in Figure 3(b) represents the transition in a given period of time.

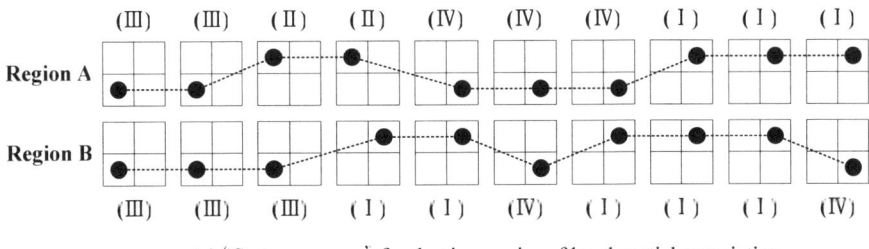

(a) "State sequence" for the time-series of local spatial association

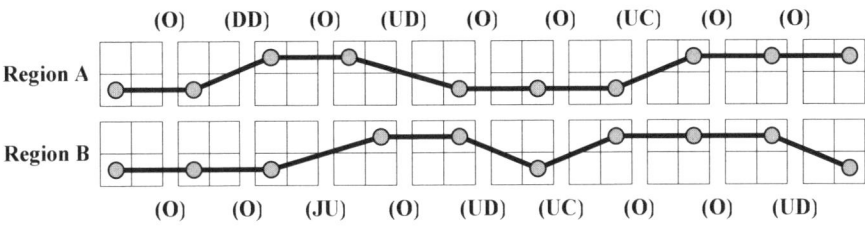

(b) "Clustering transition" for the time-series of local spatial association

Fig. 3. Comparison of the time-series signatures of local spatial association

3.3 Clustered Classification of Changing Regions

The clustered classification of changing regions in terms of the time-series of local spatial association is conducted using the similarity matrix of "state sequence" and

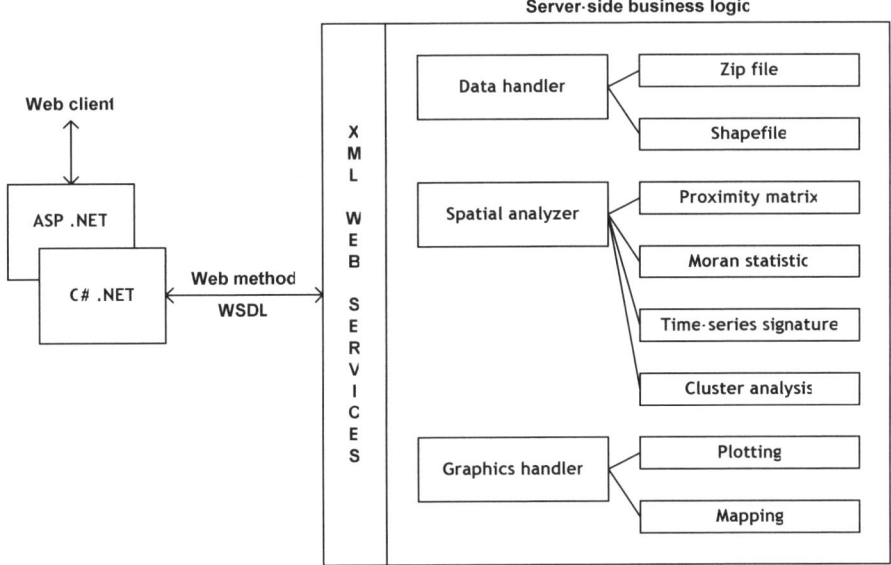

Fig. 4. System architecture based on XML Web Services

"clustering transition" of the time-series signatures. For this cluster analysis, we employ Ward method [30] that minimizes the sum of squares of any two (hypothetical) clusters.

3.4 System Architecture of Web-Based Statistical GIS

The analytical functionalities of cluster analysis on the time-series of local spatial association are implemented in the modules of server-side business logic, such as data handler, spatial analyzer, and graphics handler. The consumption of serve-side modules is carried out by the brokerage of XML Web Services that provide an open interface for Web clients in the form of a Web method. Client-side Web pages are composed in ASP .NET with the code-behind in C#. NET. The system architecture is illustrated in Figure 4.

4 Experimental Analysis

A case study of cluster analysis for the time-series signature of local spatial association is conducted by use of client-provided dataset in the form of a zip file including ESRI Shapefile. In the experimental analysis, we use the elderly population ratio of 65 administrative units in Seoul Metropolitan Area, 1995–2004 on a yearly basis. This region shows a peculiar spatial structure: its population reaches almost 50% of the total population whereas its area is approximately 10% of Korea. Moreover, since the average life span of Korean people has been elongated up to 78.2 in 2005, the analysis on elderly population ratio of Seoul Metropolitan Area might be an interesting example.

4.1 State Sequence of Time-Series Local Spatial Association

From the time-series signature of local spatial association, the accumulated "state sequence" of each region is extracted for cluster analysis. Each cluster in Figure 5 classified by Ward method seems to have experienced similar history in the state change of local spatial association. Cluster 1 is characterized by the legacy urban areas in Seoul and the rural areas in Gyeonggi where elderly population is relatively high. While cluster 2 includes the developing urban and suburban areas where elderly population is relatively low, cluster 3 has no remarkable characteristics.

4.2 Clustering Transition of Time-Series Local Spatial Association

While the analysis of "state sequence" derived from time-series signature of local spatial association is based on individual region's state change in relation to its neighbors, the analysis of "clustering transition" is about the change in clustering tendency in relation to its neighbors, such as upward/downward clustering and declustering. From the "clustering transition" of time-series signature, three clusters are classified as in Figure 6. Cluster 1 represents the regions of no change; cluster 2 includes the regions of upward change; and cluster 3 includes the regions of downward change in the clustering tendency.

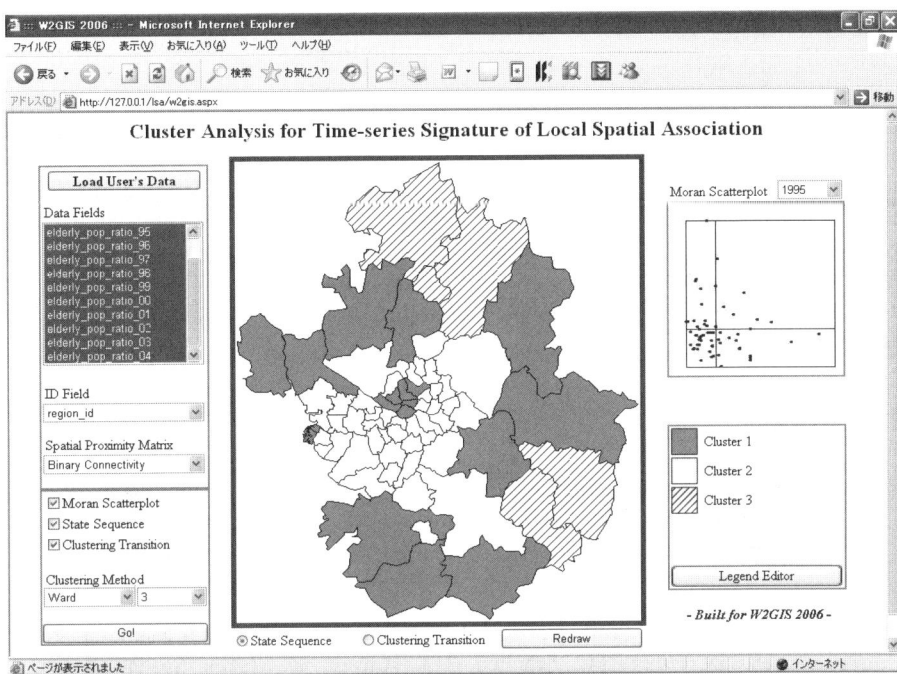

Fig. 5. Cluster analysis (I): "state sequence" of time-series local spatial association

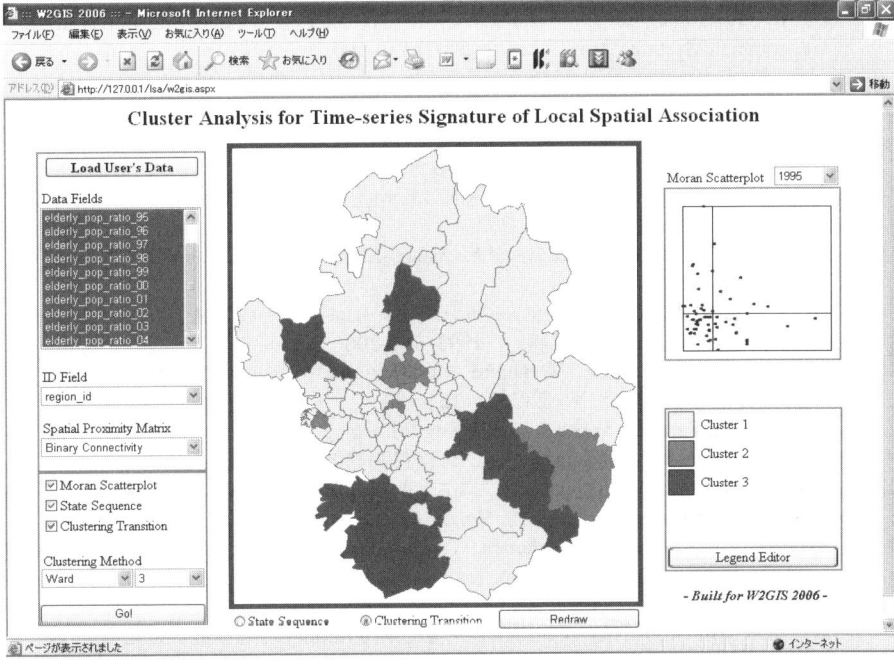

Fig. 6. Cluster analysis (II): "clustering transition" of time-series local spatial association

5 Concluding Remarks

In this paper, we focused on the concept of spatial association, one of the core philosophies in geography. In order to capture the temporal context of spatial association at local scale, we proposed a method of cluster analysis on the time-series signature of Moran scatterplot, covering "state sequence" and "clustering transition" of local spatial association. From the similarity matrix of the time-series signatures, we generated a map of the clustered classification of changing regions. By the implementation of a Web-based statistical GIS, the analytical functionalities are extended to utilizing client-provided dataset for more open access to spatial data analysis on the Web.

References

1. Andrienko, G., Andrienko, N., Voss, H., Carter, J.: Internet Mapping for Dissemination of Statistical Information. Computers, Environment and Urban Systems, Vol. 23, No. 6 (1999) 425-441
2. Anselin, L.: Local Indicators of Spatial Association - LISA. Geographical Analysis, Vol. 27, No. 2 (1995) 93-115
3. Anselin, L.: The Moran Scatterplot as an ESDA Tool to Assess Local Instability in Spatial Association. In Fisher, M., Scholten, H.J., Unwin, D. (Eds.): Spatial Analytical Perspectives on GIS, Taylor & Francis, London (1996) 111-125
4. Anselin, L., Kim, Y.-W, Syabri, I.: Web-based Analytical Tools for the Exploration of Spatial Data. Journal of Geographical Systems, Vol. 6, No. 2 (2004) 197-218
5. Boots, B.: Developing Local Measures of Spatial Association for Categorical Data. Journal of Geographical Systems, Vol. 5, No. 2 (2003) 139-160
6. Dragićević, S.: The Potential of Web-based GIS. Journal of Geographical Systems, Vol. 6, No. 2 (2004) 79-81
7. Fotheringham, A.S.: Trends in Quantitative Methods I: Stressing the Local. Progress in Human Geography, Vol. 21, No. 1 (1997) 88-96
8. Fotheringham, A.S., Brunsdon, C.: Local Forms of Spatial Analysis. Geographical Analysis, Vol. 31, No. 4 (1999) 340-358
9. Geary, R.C.: The Contiguity Ratio and Statistical Mapping. Incorporated Statistician, Vol. 5, No. 3 (1954) 115-145
10. Getis, A., Ord, J.K.: The Analysis of Spatial Association by Use of Distance Statistics. Geographical Analysis, Vol. 24, No. 3 (1992) 186-206
11. Huang, B., Worboys, M.F.: Dynamic Modeling and Visualization on the Internet. Transactions in GIS, Vol. 5, No. 2 (2001) 131-139
12. Hyyrö, H.: Bit-Parallel Approximate String Matching Algorithms with Transposition. Lecture Notes in Computer Science, Vol. 2857 (2003) 95-107
13. Jankowski, P., Stasik, M., Jankowska, M.A.: A Map Browser for an Internet-based GIS Data Repository. Transactions in GIS, Vol. 5, No. 1 (2001) 5-18
14. Kähkonen, J., Lehto, L., Kiolpeläinen, T., Sarjakoski, T.: Interactive Visualization of Geographical Objects on the Internet. International Journal of Geographical Information Science, Vol. 13, No. 4 (1999) 429-438
15. Kraak, M.-J., Brown, A.: Web Cartography. Taylor & Francis, London, UK (2001)
16. Kraak, M.-J.: The Role of the Map in a Web-GIS Environment. Journal of Geographical Systems, Vol. 6, No. 2 (2004) 83-93

17. Levenshtein, V.I.: Binary Codes Capable of Correcting Deletions, Insertions, and Reversals. Soviet Physics-Doklandy, Vol. 10, No. 8 (1966) 707-710
18. Miller, H.J.: Tobler's First Law and Spatial Analysis. Annals of the Association of American Geographers, Vol. 94, No. 2 (2004) 284-289
19. Moran, P.: The Interpretation of Statistical Maps. Journal of Royal Statistical Society, Vol. 10, No. 2 (1948) 243-251
20. Ord, J.K., Getis, A.. Local Spatial Autocorrelation Statistics: Distribution Issues and an Application. Geographical Analysis, Vol. 27, No. 4 (1995) 286-306
21. Park, K.-H., Lee, Y.-W.: A Study on Statistical GIS for Regional Analysis. Journal of the GIS Academy of Korea, Vol. 9, No. 2 (2001) 239-261
22. Peng, Z.: An Assessment Framework for the Development of Internet GIS. Environment and Planning B, Vol. 26, No. 1 (1999) 117-132
23. Rey, S.J.: Spatial Empirics for Regional Economic Growth and Convergence. Geographical Analysis, Vol. 33, No. 3 (2001) 195-214
24. Sakamoto, A., Fukui, H.: Development and Application of a Livable Environment Evaluation Support System Using Web GIS. Journal of Geographical Systems, Vol. 6, No. 2 (2004) 175-195
25. Takatsuka, M., Gahegan, M.: Sharing Exploratory Geospatial Analysis and Decision Making Using GeoVISTA Studio: From a Desktop to the Web. Journal of Geographic Information and Decision Analysis, Vol. 5, No. 2 (2001) 129-139
26. Takatsuka, M., Gahegan, M.: GeoVISTA Studio: A Codeless Visual Programming Environment for Geoscientific Data Analysis and Visualization. Computers and Geosciences, Vol. 28, No. 10 (2002) 1131-1141
27. Tobler, W.: A Computer Movie Simulating Urban Growth in the Detroit Region. Economic Geography, Vol. 46, No. 2 (1970) 234-240
28. Tsou, M.-H., Buttenfield, B.: A Dynamic Architecture for Distributing Geographic Information Services. Transactions in GIS, Vol. 6, No. 4 (2002) 355-381
29. Tsou, M.-H.: Integrating Web-based GIS and Image Processing Tools for Environmental Monitoring and Natural Resource Management. Journal of Geographical Systems, Vol. 6, No. 2 (2004) 155-174
30. Ward, J.: Hierarchical Grouping to Optimize an Objective Function. Journal of the American Statistical Association, Vol. 58, No. 301 (1963) 236-244

A Methodology for Predicting Performances of Map-Matching Algorithms

Hassan A. Karimi, Thomas Conahan, and Duangduen Roongpiboonsopit

Geoinformatics Laboratory, School of Information Science and Telecommunications,
University of Pittsburgh
135 N. Bellefield Avenue, Pittsburgh, PA, 15260, US
hkarimi@mail.sis.pitt.edu, tjc1@pitt.edu, dur2@pitt.edu

Abstract. Map matching is not always perfect and sometimes produces mis-matches. Thus, there is a degree of uncertainty for how well a map-matching algorithm will perform under certain circumstances. Circumstantial factors include accuracies of sensor data and surrounding road network structure, among others. This paper attempts to shed light on this uncertainty and proposes a methodology for predicting performances of map matching algorithms at given locations on a digital road network. In short, using a vehicle's position, the proposed methodology can be employed to predict the performance of a map-matching algorithm at that position. Since map-matching algorithms are differ-ent in their logic of matching vehicle's positions to road segments, there should be a separate prediction algorithm based on the methodology for each map-matching algorithm. To demonstrate the methodology's benefits, a probability algorithm to predict the performance of a point-to-curve map-matching algo-rithm is outlined.

Keywords: Map-Matching, Prediction, Probability, Vehicle Navigation System.

1 Introduction

Today vehicle navigation systems are becoming increasingly popular as technology im-proves and prices decline. These systems provide turn-by-turn directions creating a safer, easier, and more efficient driving experience. They also inform users of their estimated position on a digital map, making them appealing to a wide variety of applica-tions. In order to know a driver's location a procedure called map matching is em-ployed, where an estimated location (from a positioning sensor like GPS) is merged to a digital map. The end result is a map-matched position on a digital road network which is calculated from the estimated position and corresponds to the vehicle's actual loca-tion. There are three key components to map matching: (1) positioning data, (2) a digital map, and (3) a map-matching algorithm [9]. Each component has the potential to introduce error and thereby cause poor matching results [9]. Essentially, map-matching algorithms (MMAs) process noisy input data and output a precise estimate of the actual position. Although the algorithm may make a match, it is uncertain how accurate the match truly is. Factors such as digital map errors, positional uncertainty, and algorithmic shortcomings create varying levels of matching accuracy.

J.D. Carswell and T. Tezuka (Eds.): W2GIS 2006, LNCS 4295, pp. 202–213, 2006.
© Springer-Verlag Berlin Heidelberg 2006

Map-matching can be broken down into three major tasks: (1) creating the problem space, (2) selecting the best road segment, and (3) projecting onto that road segment. These tasks are the unique set of rules that define and govern the operational procedure of the algorithm and must be accomplished in order to complete the matching process.

Because map-matching is an estimation, it sometimes provides inaccurate matches or "mismatches". Currently, there is no method to guarantee the accuracy of a map-matched position but having this information could greatly benefit a number of applications. Being able to predict map-matching performance allows applications to make better decisions based on this knowledge, potentially enhancing efficiency and productivity. For example, when planning a route for a vehicle, high matching performance may be desired whereby the vehicle is very likely to be located at the matched position. Forehand knowledge of the map-matching accuracy can assist in choosing a route with the desired matching performance. Other uses include assisting MMAs overcome ambiguous matching situations and improving vehicle-to-vehicle (V2V) communication by enhancing positioning knowledge [2].

In this paper a methodology for predicting the performance of a MMA is presented. For a particular matching scenario, where the vehicle's location is known and a specific MMA is given, the predictive methodology is able to indicate the probability that the given algorithm will make an accurate match. The basic idea is to run a simulation of the algorithm at a desired location on a digital road network in order to predict the algorithm's performance. Consequently, it is possible to determine the error distribution of the sensor data at this location and thus indicate which points within the distribution will match correctly and which will not. Knowing the matching probability will indicate the success rate of the algorithm and is accomplished by coupling the probability density distribution of positioning data with the logic of a particular MMA.

The goal of this methodology is to provide users with the level of confidence of the accuracy for map-matched positions. In Section 2 the details of the methodology are presented, starting with the needed inputs and continuing with each subsequent process of the solution. In Section 3 a map matching probability algorithm (MMPA) for a point-to-curve MMA is described. The MMPA demonstrates how the performance of the given point-to-curve algorithm could be predicted. In Section 4, conclusions and future research are discussed.

2   Methodology for Predicting the Performances of MMAs

The methodology for predicting the performance of a MMA is to create a MMPA that emulates the MMA's matching procedure. Examining the general structure of a MMA reveals some common trends which include a method for constructing a problem space, a selection process for selecting the best road segment, and a projection process for projecting onto the selected road segment. These steps are the governing rules that define the MMA.The goal of the MMPA is to recreate these steps in such a way as to output a probability or level of confidence for the given MMA.

MMAs are different from MMPAs in two important aspects. First, they differ in the type and content of the information received as input, and secondly, they differ in their final goal and information output. For example, a MMA does not know the vehicle's actual location and instead uses a known position point (obtained through GPS receivers and/or some other positioning sensors) to estimate the vehicle's location. In this case, the input information is a discrete positioning point and is used to estimate the location of the vehicle. Conversely, the MMPA starts with the vehicle's true location and determines the matching performance by examining a probability density distribution of the positioning points. Here, positional data can fall anywhere within a certain range of the vehicle's actual location and is known as the sensor error region. The MMPA determines what positioning points within the sensor error region will cause the MMA to match to the correct location (the vehicle's true position). This then indicates the success rate for that MMA at that particular vehicle location.

Figure 1 represents the general methodology for finding the level of confidence for any MMA. In essence, the methodology consists of the general set of tasks for the MMPA. An explanation of the input and each of the steps will follow and an example algorithm is then detailed.

2.1 Inputs to MMPA

In Figure 1 the inputs to MMPA are contained within the dashed box and consist of the road network, the vehicle's positions, the algorithm logic, and the sensor error. The given road network is used as the sample set of road segments for the map-matching process. The given vehicle positions are the known locations of the vehicle on the road network. The given algorithmic logic provides the rules for constructing the problem space, road segment selection, and position projection processes. The given sensor accuracy is the range of error representing the probable location of the sensor data. Each of these plays an important role in understanding the matching performance and each is detailed in Figure 1.

2.1.1 Road Network

The road network is the vector representation of the set of roads (N) for a given area. Each of the roads within N is represented as a piecewise linear set of curves in $\mathbf{R}^{2}$, called a road segment. Each road segment is created by a series of shape points and terminated by a node [5]. However, because a vector road network is a representation, it will inevitably contain errors. In [9], an explanation detailing map errors and how they impact the map matching process is discussed. Some common errors include missing roads, multilane roads represented as a single centerline, the lack of height information on a 2D surface, and inaccurate road geometries.

In the methodology presented here some of these errors will have no impact on the predictive performance. For example, since the MMPA is conducted at a known position on a digital road network, the known location must be on a road segment and cannot be on a road that is not present in the road network. Therefore, missing roads will not affect the MMPA because all probabilities will be calculated at positions on existing roads within the network. However, from the MMPA's point of view, inaccurate road geometries and the lack of height information will seem to impact sensor data, and not the surrounding road structure. Further research may be needed to gauge the impact of such inaccuracies.

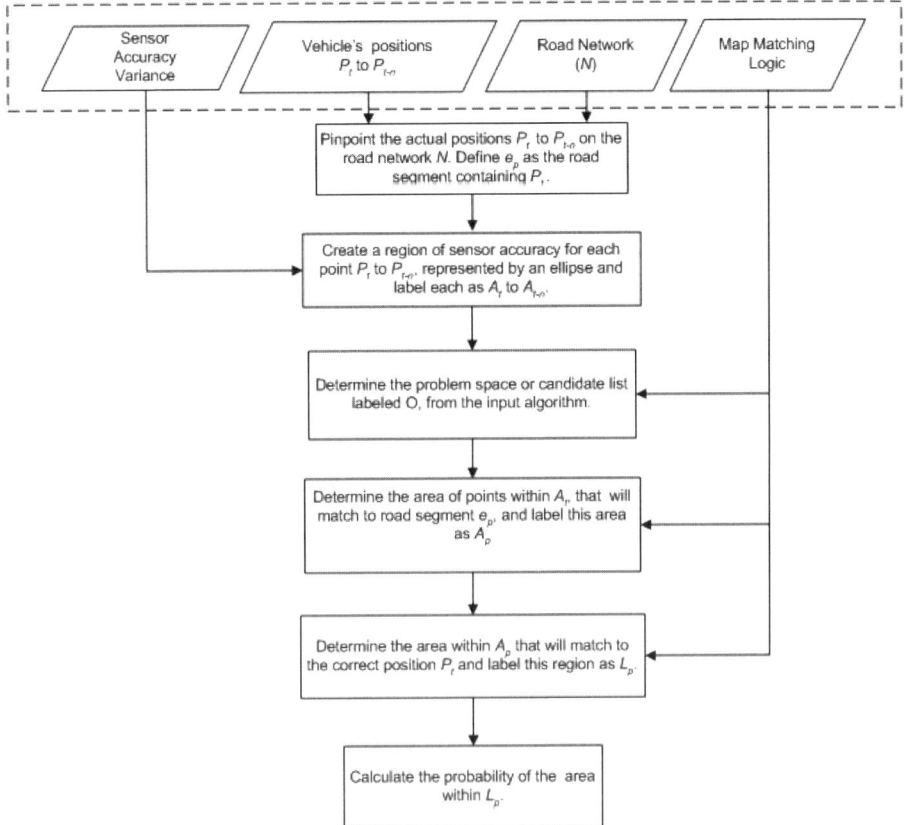

Fig. 1. Concept of the predictive map-matching methodology

2.1.2 Vehicular Positions

The vehicle's positions are the known locations of a vehicle on the road segments at times t to t-n within the digital road network. These positions are labeled P_t to P_{t-n} and are input by the user in accordance with the requirements of the given MMA. For example, given a simple point-to-curve MMA that does not use any historical data for its computations, the location P_t represents the known location of the vehicle at time t, is the only historical point needed, and is the position at which the performance of the algorithm is measured. However, other MMAs may require historical position points. In another example, finding the performance of a curve-to-curve MMA requires a position P_t as well as subsequent positions P_{t-1} to P_{t-n} with n depending on the requirements of the algorithm. The basic premise of a curve-to-curve algorithm uses the previous n number of points to construct an imaginary road segment to be compared to the surrounding road segments within the problem space. In order to successfully complete this task, the MMPA will need to recreate the imaginary road segments from points P_t to P_{t-n}.

Generally speaking, the vehicle's positions are the locations on the road network where, in a real world situation, the vehicle would be updated with positioning data. It is important to note that this is not the estimated position received, for example, by a GPS unit, but rather the known position of the vehicle where the GPS unit received signals from the satellites. There must be at least one known position P_t and any number of others as required by the MMA. P_t represents the most recent vehicle position and is the location where the MMPA finds the final probability.

2.1.3 Map-Matching Logic

The algorithmic logic is considered to be the set of steps followed by a MMA to achieve the three tasks of the matching process: defining the problem space, selecting the correct road segment, and projecting a position onto the selected road segment. The MMPA models the logic of the MMA. For example, using a variation of the point-to-curve algorithm in [5], where heading and topological information are used, the problem space is constructed through topological connectivity. In other words, all road segments connected to the most recently matched road segment are included in the problem space. Then, the road segment within the candidate list that has the shortest distance to the input GPS point is chosen. Finally, the MMA projects the GPS point to the closest location on the selected road segment. The MMPA would use the rules outlined by the MMA above, to reconstruct the matching scenario for the probability distribution area of the GPS points.

2.1.4 Sensor Accuracy

The sensor accuracy input is the probability distribution area for the vehicle's positioning data. Different positioning sensors have differing levels of accuracy. From these sensors, a range of error can be constructed outlining the probability distribution in which a positioning point will fall. This area is considered to be the range of error variance, labeled A_t through A_{t-n}. As mentioned in sub-section 2.1.2, these error ranges are constructed for the corresponding vehicle's positions P_t to P_{t-n}. In other words, the vehicle's positions will then represent the positions of error variance of the sensor data.

2.2 Processing Steps

In Figure 1 there are six steps outlining the process needed to determine the probability. These steps use the input described above and are designed to mimic the general[1] map matching process. Steps 1 and 2 are used for initialization, Step 3 defines the problem space, Step 4 selects the road segment, Step 5 selects the position, and Step 6 is used to calculate the final probability. Each of these steps is reviewed in detail.

2.2.1 Steps 1 and 2: Initialization

The methodology works on the principle of recreating a map-matching scenario for a specific algorithm using a continuous probability density function (PDF) as the positioning input. The first two steps of the process map the PDF to the digital road network. These steps will have the same general procedure for all MMPAs but will differ in details like the number of vehicle positions used and error ellipses constructed.

---

[1] The word general is used here to indicate the unspecified map matching process. When using this methodology to construct an algorithm the general map matching process will be replaced with the process defined by a given MMA.

Step 1 simply overlays the known vehicle's positions, P_t through P_{t-n}, onto the digital road network N. The second step follows by overlaying the error region A_t through A_{t-n} for each position point P. The road segment that contains P_t is labeled as e_p.

2.2.2 Steps 3: Defining the Problem Space

This step determines the set of candidate road segments according to the procedure outlined by the given MMA. There are many different approaches to determine the problem space and some of the most common use a method of topology [3][5] or a range of error variance [4][8]. Careful attention is needed to determine the problem space so as not to introduce error. If the MMA's matching process is not successfully replicated in the design of the MMPA, the output probability of the MMPA will not be accurate. As an example, MMAs such as those described in [4][8], determine the problem space by constructing an ellipse of error variance around the input GPS point. The road segments that fall within this ellipse are considered as the list of candidate road segments. Each respective algorithm then performs a unique set of procedures on the problem space to determine the road segment with the best[2] match. The MMPA attempts to imitate this procedure using a continuous probability distribution instead of an exact position point as input. The MMPA creates the problem space according to the logic of the MMA and on a region-by-region basis within the sensor error ellipse A_t. This means that there could be many different problem spaces within one error ellipse A_t, with each position "seeing" its own set of candidate road segments. The problem space is constructed on a region-by-region basis as opposed to one large, communal problem space because different regions within the ellipse A_t may have a specific set of road segments that influence the matching process that are unique to that region.

2.2.3 Steps 4: Choosing Road Segment

The goal of this step is to determine the percentage of all points within A_t that will select the known road segment e_p which contains P_t. Using the problem space(s) determined in the last step, the MMPA recreates the selection process and maps out the geometric area of points that select e_p from the problem space and labels this area as A_p.

There are many different MMAs and hence many different ways of completing this task. The complexity of this step can range from the very simple, as seen with algorithms that choose the closest road segment [5], to the very complex, as seen with algorithms that use fuzzy logic systems [1]. Recreating the selection process in terms of probabilities is necessarily more complex than the map-matching situation being modeled. Careful attention should be given to the MMPA when recreating this step. If the selection process is not recreated accurately error will be introduced into the final probability.

2.2.4 Steps 5 and 6: Projecting Position and Calculating the Probability

The goal of this step is to determine the region of points within A_p that correctly project to position P_t. Running a simulation of the points within A_p using the logic of a given MMA will output the area of points that will map to the correct position P_t. This area is labeled L_p and represents all of the points within A_p that will match to the

---

[2] The road segment within the problem space that most represents the vehicles actual location.

correct road segment e_p and position P_t, giving a successful match. The final step of this methodology is to determine the probability of the region within L_p. This will indicate the level of confidence for a given MMA at a known position on a digital map.

3 Example Scenario

In this section, a MMPA to determine the level of confidence for a modified version of the point-to-curve algorithm described in [5] is discussed. This algorithm uses GPS to receive positional data. The algorithm begins by establishing a problem space using a range query around the GPS point. Then the best road segment is chosen through a method of shortest distance. Finally, the map-matched position is projected onto the chosen road segment at the point of shortest distance. In our methodology, a MMPA is developed to mimic the map matching procedures for this algorithm. Figure 2 represents the flowchart illustrating how the MMPA works.

The point-to-curve algorithm was chosen because of its simplicity and instructional facilitation. However, most MMAs are much more complex and include other criteria such as heading information, speed, and topological connectivity, among others.

3.1 Initialization

Let P_t represents a point consisting of a pair of coordinates (x_t, y_t) at time t. This point will indicate the vehicle's position on the road segment. First, P_t is first located on a road segment e_p in the road network N. We assume that the road network N is reasonably accurate and P_t is located on a road segment. It is necessary for the MMPA to know the actual position of the vehicle on the road segment because the MMPA will calculate the performance of the MMA at position P_t. If the vehicle's actual position is not near P_t then the calculated performance will not reflect the vehicle's actual position and be rendered useless.

After locating P_t, the next step is to create a confidence region around P_t labeled A_t. This confidence region A_t is created by the predicted sensor variance-covariance of 2D coordinate system (x, y). Predicting this variance-covariance is beyond the scope of this paper. However, there is a 99% probability that the positioning points received from the GPS will fall within the bounds of A_t. An error ellipse is chosen to represent this region with the length of the semi-major axis, a, and the length of the semi-minor, b, and the orientation of the semi-major axis relative to North, ϕ. To obtain a 99% confidence level, an expansion factor $\hat{\sigma}_0$ of 3.03 is used [10].

$$a = \hat{\sigma}_0 \sqrt{\frac{1}{2}\left(\sigma_x^2 + \sigma_y^2 + \sqrt{\left(\sigma_x^2 - \sigma_y^2\right)^2 + 4\sigma_{xy}^2}\right)} \tag{1}$$

$$b = \hat{\sigma}_0 \sqrt{\frac{1}{2}\left(\sigma_x^2 + \sigma_y^2 - \sqrt{\left(\sigma_x^2 - \sigma_y^2\right)^2 + 4\sigma_{xy}^2}\right)} \tag{2}$$

$$\phi = \frac{\pi}{2} - \frac{1}{2} \arctan \left(\frac{2\sigma_{xy}}{\sigma_x^2 - \sigma_y^2} \right) \tag{3}$$

where σ_x and σ_y are the variances of the sensor measurement errors, and σ_{xy} is the covariance.

3.1.1 Defining the Problem Space

After P_t is located and its corresponding confidence region A_t established, then the problem space can be calculated. The rule set outlined by the given map matching logic governs how the probability algorithm will choose the problem space, select the best road segment, and project onto the selected road segment. In the algorithm presented in [5] the problem space is defined using a range query by selecting all the road segments within a calculated range around the GPS point. Because this algorithm does not use a complex method for determining the best road segment (for the second step of the matching process), we can make one large problem space for all points within the confidence region A_t. This problem space will include all the road segments that fall within an area that is twice the size of the confidence region A_t. This is possible because this MMA uses the method of shortest distance to calculate the best road segment. The extra road segments introduced through the communal problem space will not change the outcome of the matching probability because the extra road segments will not influence the shortest distance.

The problem space O_t is plotted as having twice the size of area (A_t) and identifies the boundary within which candidate road segments are chosen for the next step. All road segments (e_i where i = 1 to n) within the problem space except the actual road segment e_p, are collected in $E = \{e_1, e_2, ..., e_n\}$. Figure 3 shows the properties of this configuration.

3.2 Road Segment Selection Process

After defining the problem space, the next step is finding the area within the confidence region A_t that matches to the road segment e_p. Since we are only concerned with finding the region of points that match to e_p, we can eliminate some road segments from set E. Accomplishing this task, begins by examining each road segment within E and determining if an unintersected line can be drawn from the selected road segment to e_p. If the line drawn from e_i intersects another road segment before it reaches e_p, road segment e_i is eliminated from set E. The remaining road segments are those road segments that will be used to calculate the area of points that match to e_p. To identify this area the MMPA constructs the median lines l_j between e_p and each of the remaining road segments in set E. Each l_j is a line segment that is equidistant from road segment e_i and road segment e_p.

Let L be the set of median lines so that $L = \{l_1, l_2, ..., l_m\}$. Due to the characteristic of N, that is, all road segments are piecewise linear, median lines are also piecewise linear. In addition, l_j indicates the ambiguous position where the point-to-curve algorithm has to choose between two shortest road segments of the same distance. Therefore, we assume that only those points that fall within the boundaries defined by L will match correctly to e_p.

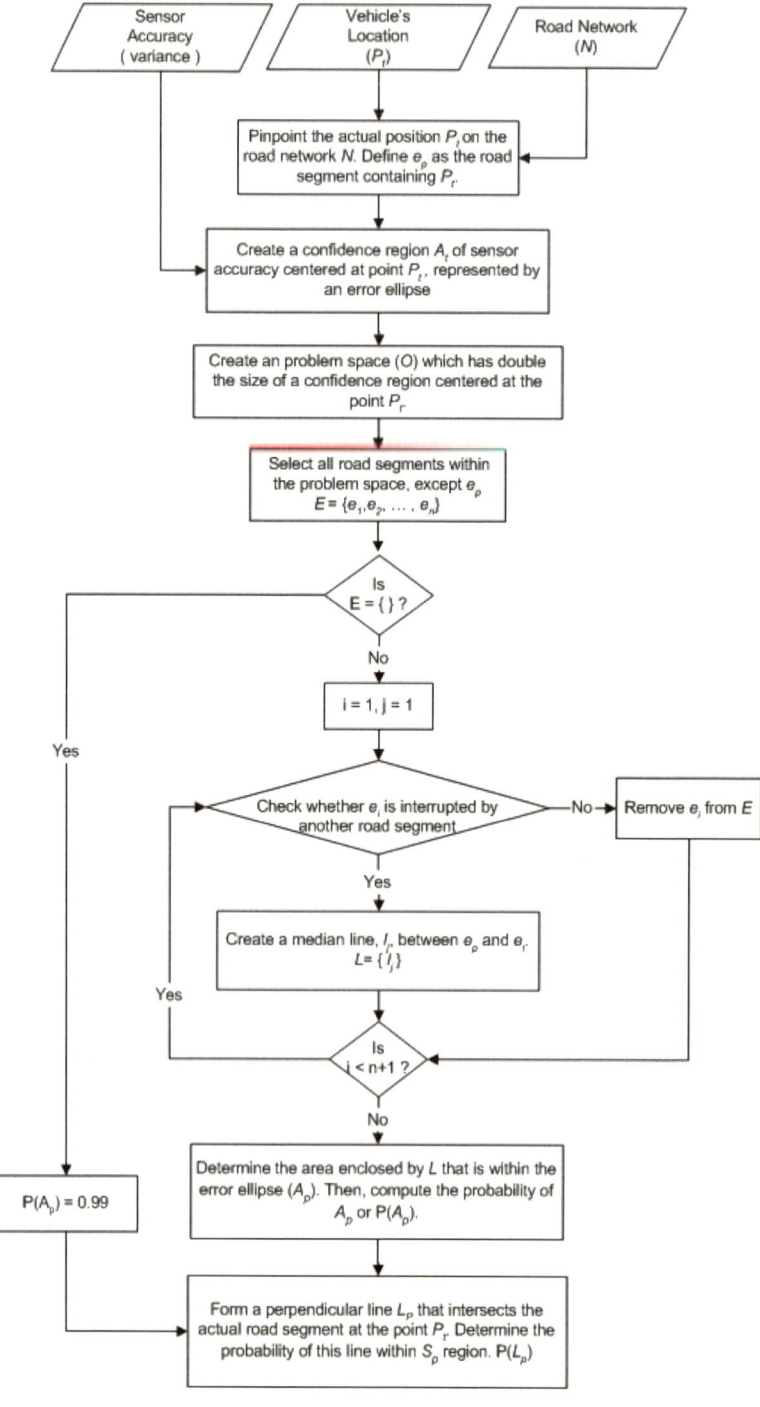

Fig. 2. Flowchart for the point-to-curve probability algorithm

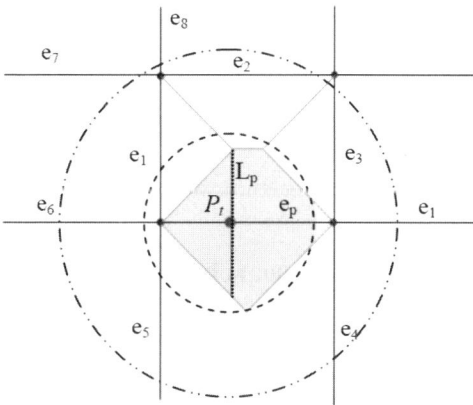

Fig. 3. Final result for the MMPA constructed for the point-to-curve MMA

3.3 Position Projection Process

The area within the error ellipse enclosed by L represents those points that match to e_p and this area is named as A_p. We define the probability of A_p that the confidence level given a correct road segment as $P(A_p)$. As mentioned in the initialization step, the confidence region A_t has the probability 0.99. Thus, if A_p is a part of the error ellipse A_t, $P(A_p)$ is less than 0.99 as presented by the shaded area in Figure 3. To be more accurate in computing $P(A_p)$, the bivariate normal distribution function is taken into account as presented in (4) [6].

$$p(x, y) = \frac{1}{2\pi\sigma_x\sigma_y\sqrt{1-\rho^2}} \exp\left[-\frac{1}{2(1-\rho^2)}\left[\left(\frac{x-\mu_x}{\sigma_x}\right)^2 + \left(\frac{y-\mu_y}{\sigma_y}\right)^2 - 2\rho\left(\frac{x-\mu_x}{\sigma_x}\right)\left(\frac{y-\mu_y}{\sigma_y}\right)\right]\right] \quad (4)$$

where $\rho = \dfrac{\sigma_{xy}^2}{\sigma_x\sigma_y}$ is the spatial correlation coefficient $-1 \le \rho \le 1$. Therefore, the

probability of the area can be computed by using (5)

$$P(A_p) = \int_{A_p} p(x, y)\,dxdy \quad (5)$$

However, if no road segment within O_t is selected to E which is processed in the previous step, all possible positioning points within the confidence region will be snapped on e_p. In other words, when $A_p = A_t$, the confidence level given the correct road segment is 0.99.

3.4 Calculating Level of Confidence

Because the point-to-curve algorithm uses the shortest distance as the criterion for projecting onto a road segment, we know that this algorithm will give a correct position only if points fall on a perpendicular line intersecting e_p at P_t. This line is defined as L_p as represented in Figure 3. Therefore, the probability of L_p is the confidence level given a correct position and can be compute by (6)

$$P(L_p) = \int_{L_p} p(x, y) dy \qquad (6)$$

where L_p is located at $x = 0$.

4 Conclusions and Future Research

In this paper, we presented a methodology for calculating the performance of a MMA at a known position on a digital road network. The methodology provides the procedures for constructing a MMPA that imitates a given MMA in order to calculate its matching probability. This is done by processing each of the MMA's component parts in terms of the error distribution region for the positioning data. The MMPA outputs the level of confidence that the MMA will make a correct match.

To demonstrate the proposed methodology, we outlined a specific MMA that employs a point-to-curve with the shortest distance matching approach. The results of our research indicate that a probability algorithm can be constructed to measure the matching performance of a MMA, thus providing users with a level of confidence on the matched positions.

The MMA used for the proposed algorithm is fairly simple and further work is needed to determine how this methodology would apply to more complex algorithms such as those that use curve-to-curve (with many historical points and therefore joint probabilities) and or other matching approaches like fuzzy logic. Further research may also examine how map errors affect the MMPA including where the known position P_t should be located on the digital map (multi lane road effect) and how geometric map errors impact the output of the MMPA. Limitations of the methodology only allow the calculations to be accurate where the vehicle's position is known.

References

1. Fu, M., Li, J., Wang, M.: A Hybrid Map Matching Algorithm Based on Fuzzy Comprehensive Judgment. In: IEEE Intelligent Transportation Systems Conference, Washington (2004)
2. Galler, B.A., Asher, H.: Vehicle-to-Vehicle Communication for Collision Avoidance and Improved Traffic Flow. ITS-IDEA Program Project Final Report. Transportation Research Board (1995)
3. Greenfeld, J.S.: Matching GPS Observations to Locations on a Digital Map. In: Proceedings of the 81th Annual Meeting of the Transportation Research Board, Washington D.C. (2002)
4. Ochieng, W. Y., Quddus, M. A., Noland, R. B.: Map-Matching in Complex Urban Road Networks. Brazilian Journal of Cartography (Revista Brasileira de zartografia), 55 (2), 1–18. (2003)
5. White, C.E., Berstein, D., Kornhauser, A.L.: Some Map Matching Algorithms for Personal Navigation Assistants. Transportation Research Part C 8 (2000) 91-108
6. Wilks, D.S.: Statistical Methods in the Atmospheric Sciences. Academic Press, 1995.

7. Yang, X., Liu, J., Zhao, F., Vaidya, N.: A Vehicle-to-Vehicle Communication Protocol for Cooperative Collision Warning. Technical Report. University of Illinois at Urbana-Champaign (2003)
8. Zewang, C., Yongrong, S., Xing, Y.: Development of an Algorithm for Car Navigation System Based on Dempster-Shafer Evidence Reasoning. In: Submitted to IEEE International Conference On Intelligent Transportation Systems, Singapore (2002)
9   Zhang, X , Wang, Q , Wan, W.: The Relationship Among Vehicle Positioning Performance, Map Quality and Sensitivities and Feasibilities of Map-Matching Algorithms. In: Intelligent Vehicles Symposium proceedings, Columbus, Ohio (2003)
10. Zhao, Y.: Vehicle Location and Navigation System. Artech House, Inc., MA. (1997)

Path Planning for Chaining Geospatial Web Services

Peng Yue[1,2], Liping Di[1,*], Wenli Yang[1], Genong Yu[1], and Peisheng Zhao[1]

[1] Center for Spatial Information Science and Systems (CSISS)
George Mason University
6301 Ivy Lane, Suite 620, Greenbelt, MD 20770
[2] State Key Laboratory of Information Engineering in Surveying, Mapping and Remote
Sensing, Wuhan University
129 Luoyu Road, Wuhan, China, 430079
Tel.: 1-301-982-0795; Fax: 1-301-345-5492
{pyue, ldi, wyang1, gyu, pzhao}@gmu.edu

Abstract. Semantic Web technologies provide a promising prospect for automatic discovery and chaining of geospatial Web services. This paper addresses semantic geospatial Web services, particularly the path planning for service chaining. We use OWL-S to represent the geospatial semantic Web service. A graph with nodes representing services and connection weights representing degrees of semantic matching between nodes is formulated using information from multiple geospatial semantic Web services. The graph is used to build logical path models, which can be instantiated to a physical service chain for execution. A prototype system, which includes a real world geospatial model, is implemented to demonstrate the concept and approach.

Keywords: Geospatial Web Service; Service Composition; Service Chaining; Path Planning; Semantic Web Service; OWL-S.

1 Introduction

Tens of thousands of highly multidisciplinary, heterogeneous, and distributed datasets are now on-line. The data volume is climbing in an ever increasing speed. NASA's Earth Observing System (EOS) alone is generating about 3.5 terabytes of data each day. This vast data source has become indispensable for many geospatial research and applications. While presenting unprecedented opportunities for people to explore, it at the same time brings substantial challenges. It is currently not possible to make fully use of this valuable data reservoir. Recent developments in Web service technologies offer promising prospective for people to significantly enhance their ability of using online/near-line data in the Web. The technologies provide interoperability among different services through standard descriptions of service interfaces. The interoperable services can be published, discovered, and accessed through the Web. A number of interoperable services have been available in the geospatial community, most notably the Open Geospatial Consortium (OGC) standard-compliant services, including Web Feature Service (WFS), Web Map Service (WMS), Web Coverage

---

* Corresponding author.

J.D. Carswell and T. Tezuka (Eds.): W2GIS 2006, LNCS 4295, pp. 214–226, 2006.

Service (WCS), Web Coordinate Transformation Service (WCTS), Web Image Classification Service (WICS), and Catalogue Services for Web (CSW).

The power of geospatial Web services is, however, not just in their interfaces, but also in their potentials of being chained into the composite service to solve complex geospatial problem. To achieve the goal of automation of Web service composition, it is necessary to make Web services semantically meaningful, in addition to syntactically expressiveness. For example, the OGC WCS interface unambiguously defined the syntax for requesting a coverage data set. It does not tell, however, how to obtain a surface temperature data set instead of a soil moisture data set. Similarly, the WICS interface defines its input being a multiple-band image, which is essentially is a three dimensional (3D) data array. This input is not different from another service also taking a 3D array as input, such as a color compositing service. Semantic description of Web service and semantic interoperability ensure that *right* services are invoked to produce *right* outcomes, as oppose to syntactic interoperability that ensure services are invoked correctly. Geospatial semantic Web services described in this paper are those geospatial services enriched with semantic descriptions using such technologies as OWL-S[1], WSMO[2], WSDL-S[3], and SWSF[4]. The properties of individual geospatial Web services, such as input, output and service functionality, are advertised with explicitly formalized semantic representation. In this way, the content transferred between the Web services becomes machine understandable, permitting automatic chaining of *right* geospatial Web services.

This paper addresses semantic description of geospatial Web services, reasoning based on service semantics, and connecting such services automatically for geospatial modeling. The emphasis is on how to find the *right* services and *optimum* paths among the services. We use OWL-S to represent the geospatial semantic Web service. A real world geospatial model is presented as the graph formulated using information from multiple geospatial semantic Web services. Nodes in the graph represent services and connectivity or edge weight is determined by the semantic matching of input and output of the services. The final optimum path is determined through path planning which consists of three interactive phases: path modeling, plan instantiation and service chain execution. The method presented in this paper can be used to answer specific geospatial-related "what if" questions in a Web service environment. A prototype system has been implemented to demonstrate the concept and approach. A use case on landslide susceptibility assessment has been employed in this online system to illustrate the applicability of this method to the real world problem.

2 A Use Case

A typical real world geospatial problem can often be represented as a "what is" or "what if" question for a certain location at a certain time, for example, "what is the landslide risk for location *L* at time *T*?".

---

[1] OWL-based Web Service Ontology. http://www.daml.org/services/owl-s/1.1/
[2] Web Service Modeling Ontology (WSMO). http://www.w3.org/Submission/WSMO/
[3] Web Service Semantics - WSDL-S,
     http://www.w3.org/2005/04/FSWS/Submissions/17/WSDL-S.htm
[4] Semantic Web Services Framework (SWSF), http://www.w3.org/Submission/2005/SUBM-SWSF-20050909/

Table 1. Services used in this example

Service	Description
Landslide Susceptibility	Two computational models for landslide susceptibility are provided. One takes into consideration the factors of terrain slope, terrain aspect, land cover types, and vegetation conditions (through Normalized Difference Vegetation Index-NDVI) by assigning each a weighting factor and then doing the map algebra computation. The other takes into consideration only the factors of terrain slope and terrain aspect with different weighting factors.
Slope	Computes the terrain slope from Digital Elevation Model (DEM) data.
Slope Aspect	Generates the terrain aspect from DEM data.
ETM NDVI	Calculates ETM NDVI based on the Near-infrared (NIR) image (i.e. ETM Band 4) and red image (i.e. ETM Band 3).
WICS	Performs the image classification functions (supervised) that can generate the land cover types.
WCS	Provides the available geospatial data in the data archives

Automating the chaining process for answering such geospatial questions is characterized by several properties:

(1) Data-centric discovery: The answer to a geospatial question is some kind of data, or more precisely, high level information or knowledge, for example, the landslide susceptibility data, Such high level information or knowledge is usually not directly available, especially for a specific location and time and thus some "service" is needed to derive them. We assume two simple computation models for landslide susceptibility index are available as *landslideSusceptibility* services: one takes consideration of terrain slope and aspect, land cover type, and vegetation growing condition, and the other is based only on terrain slope and aspect. In each of the computation model, it also involves other models, such as deriving terrain slope from digital terrain model (DEM) and calculating Normalized Difference Vegetation Index (NDVI) as an indicator of vegetation growing condition. These other models may also involve more models. When both services and data/information/knowledge can be correctly described based on their thematic meanings and such descriptions are advertised in widely accessible catalogues, the answer to a particular geospatial question is potentially always available through reasoning on the thematic descriptions of data/information and recursively call related services for those data/information. This is a data-driven backward chaining process that creates an executable service chain starting with available input data and ending at the answer to a question.

(2) Semantic interoperability: Interoperability is the capability to exchange information, execute programs, or transfer data among various functional units in a manner that requires the user to have little or no knowledge of the unique characteristics and internal implementation of those units [1]. Semantic interoperability assures that the contents of data and services are correctly understood when data/services are connected.

For example, a catalogue may show availability of an "*ETM NDVI*" service that can produce the "*ETM NDVI*" type of data but no service or data for *NDVI* is available. A data-driven chaining based on exact matching will not be able to replace *NDVI* with *ETM NDVI*. Thus, there should be a knowledge representation mechanism that will let the machine understand that *ETM NDVI* data can satisfy the requirement for *NDVI*. Ontology is introduced for such representation. Ontology is a formal, explicit specification of a conceptualization that provides a common vocabulary for a knowledge domain and defines the meaning of the terms and the relations between them [2]. The Web Ontology Language (OWL) [3], recommended by W3C as the standard Web ontology language, is used as a vehicle for the knowledge representation.

Table 1 lists the services used to answer the example "what is the landslide risk for location *L* at time *T*". This example will serve the use case in our paper.

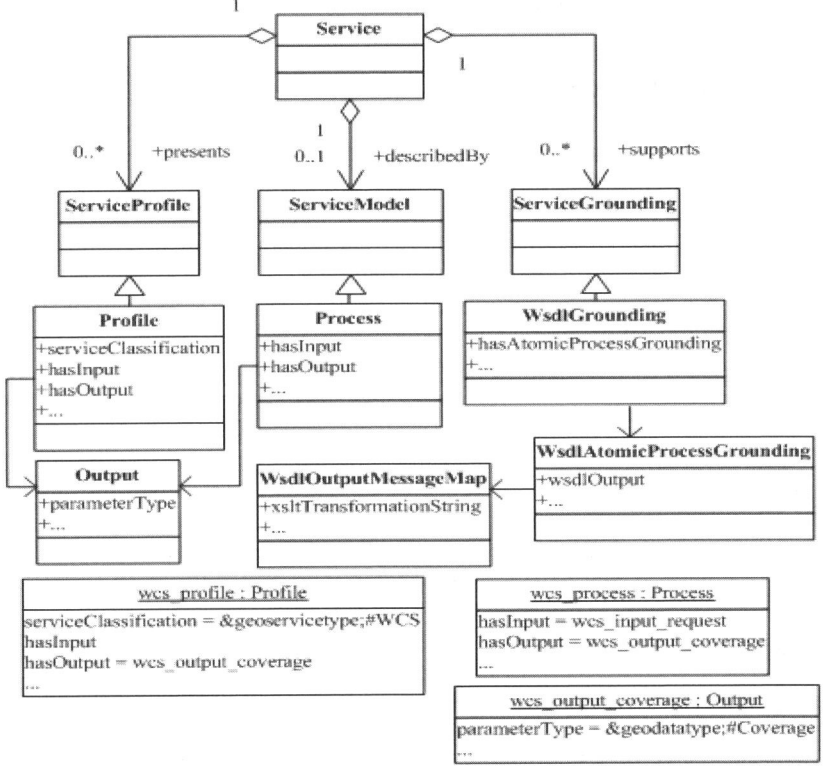

Fig. 1. OWL-S structure

3 Geospatial Semantic Web Service

As mentioned above, OWL-S is used to describe geospatial semantic Web services. Figure 1 presents a UML-style diagram illustrating how to describe WCS using OWL-S.

OWL-S consists of three main parts (Fig. 1):

(1)   service profile: what a service does (advertisement), e.g., WCS has "WCS" as its "ServiceType" and "Coverage" as its output "DataType".

(2)   service model: how a service works (detailed description), e.g., WCS has a series of input parameters which are identified in the service model.

(3)   service grounding: how to assess a service (execution), e.g., the WCS output "DataType" is grounded to the output message of the GetCoverage operation (defined in the WCS WSDL) using an XSLT transformation.

"DataType" and "ServiceType" are defined using ontologies that describe the input/output and functionality of service in OWL-S. The service profile and service model are the semantic description of the Web service. Service grounding is the relation of the semantic description to the syntactic description of a service. WSDL is generally regarded as the syntactic description.

4   Path Planning for Chaining Geospatial Web Services

A "path" is an ordered sequence of services that, when composed, can generate an executable service chain for problem solving. Thus the process of chaining geospatial Web services is a path planning process. We used a three-phase approach for the path planning [4]. The first phase is to construct a logical model in which the most suitable service types are identified and logically connected. We refer to this as modeling phase. The second phase is to generate an executable service chain, a physical model, from the logical model through finding service instances of the chained service types. We refer to this as instantiation phase. The third phase is to actually execute the service chain. This three-phase approach can be identified as an offline planning approach in Artificial Intelligence (AI) planning. Compared to the online planning where different phases are not differentiated, offline planning is usually more suitable for geospatial Web service due to the nature of geospatial problem. Most geospatial models, such as the aforementioned landslide risk models, involve complex, data- and computing-intensive processes. Given the resources usually consumed by geospatial processing services, offline planning can bring more predictability and efficiency.

4.1   Service Graph

Figure 2 shows a directed graph describing a partial landslide model. The nodes in this graph are services that will be needed to derive susceptibility index. The services are connected based on the semantic matches of their inputs and outputs, which are described by a "DataType" ontology defining the semantics of data and their hierarchical relationship. There are often multiple possible paths to research a specific node. For example, there are two landslideSusceptbility service nodes both generating landslide susceptibility index but taking different inputs. The connections between the services are assigned with positive weight values reflecting the different levels of the semantic matches. The weights are determined based on the hierarchical ontology relationships. Four levels of service matches are adopted: EXACT, SUBSUME, RELAXED and FAILED. To determine the connectivity from service *Node1* to service *Node2*, let *OntR* denote the input "DataType" of *Node2* and *OntP* denote the output "DataType" of *Node1*. The four levels of matching can be expressed as following with the increasing weight values:

EXACT: *OntR* equivalent to *OntP (Edge Weight Value=1)*
SUBSUME: *OntP* is a subclassOf *OntR (Edge Weight Value=2)*
RELAXED: *OntR* is a subclassOf *OntP (Edge Weight Value=3)*
FAILED: None of above matches. (*no connection or Edge Weight Value=+∞*)

Information needed to generate the service graph is provided by the service profile of each service's OWL-S. The graph generated from service profiles is an abstract model which does not include information about physical availability of the involved services because OWL-S does not need to specify its service groundings.

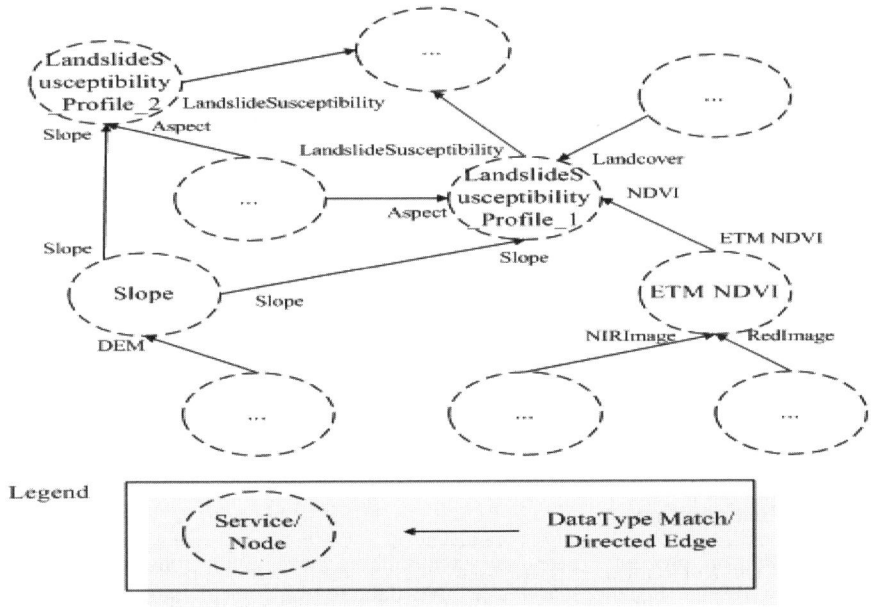

Fig. 2. A section of service graph

A section of graph is shown in figure 2. The graph is a directed graph. Let $TD(v)$ denote the node degree of node v which has m inputs, $IP = \{ip_1, ip_2, \cdots, ip_m\}$, and n outputs, $OP = \{op_1, op_2, \cdots, op_n\}$. For each input ip_i, the collection of services that can be potentially chained to v is given by $\sum_{j=1}^{i_l} S_j(ip_i)$, where i_l is the number of services the can provide semantically matched output for the ith input of node v. For each output op_i, the collection of services to which v can be chained is

$\sum_{j=1}^{i_k} S_j(op_i)$, where i_k is the number of services whose inputs semantically match the

ith output of node v.

Let $OD(v)$ and $ID(v)$ denotes, respectively, the outdegree and the indegree of node v. The node degree in the graph can be represented using equation (1).

$$TD(v) = OD(v) + ID(v) \tag{1}$$

where: $OD(v) = num\left(\sum_{i=1}^{n}\sum_{j=1}^{i_k} S_j(op_i)\right), ID(v) = num\left(\sum_{i=1}^{m}\sum_{j=1}^{i_l} S_j(ip_i)\right)$

4.2 Path Modeling

During this phase the service graph is used to find one or more sequences, or logical paths, of services whose input and output match. Each path provides a logical solution to a real world geospatial problem, e.g., landslide susceptibility in our case. The choice among various paths is subject to semantic control and various performance criteria. Semantic control includes both the correctness of a path and the degree of matching between connected services. Performance criteria are usually more important in the next phase - the plan instantiation, yet, in the current phase, it can still be used to help select a plan based on the length of the logical path. Multiple paths are usually constructed to provide alternative plans to deal with different instantiation and runtime possibilities. For example, if a required data or service is found to be not available or a service returns an error when executing a plan, the next suitable plan can be used. Considering a "what if" question, for example, "what is the landslide risk for location L at time T if vegetation were changed?", a logical path in figure 3 without the dotted rectangle is found.

4.3 Plan Instantiation

The instantiation process creates an executable service chain (physical path) by binding the service instances and available data to the logical path (i.e. plan). It consists of two steps: leaf node instantiation and service instance selection.

4.3.1 Leaf Node Instantiation

The concept of the "leaf node" is introduced to describe a service node for which at least one of its inputs, each described a specific "DataType" in the "DataType" ontology, is not connected to a service node in the logical path and thus data with correct DataType is needed for such an input. The process of binding a "DataType" to the available data is called leaf node instantiation. The available data may either be readily obtainable from some data provider or needs to be generated at run-time through a service chain. A geospatial catalogue is involved in this process to provide the information of data availability. In addition to the "DataType" constraint, more filtering requirements, such as spatial and temporal extents and data format as instantiation parameters, are added to the query on the catalogue. If the requested data

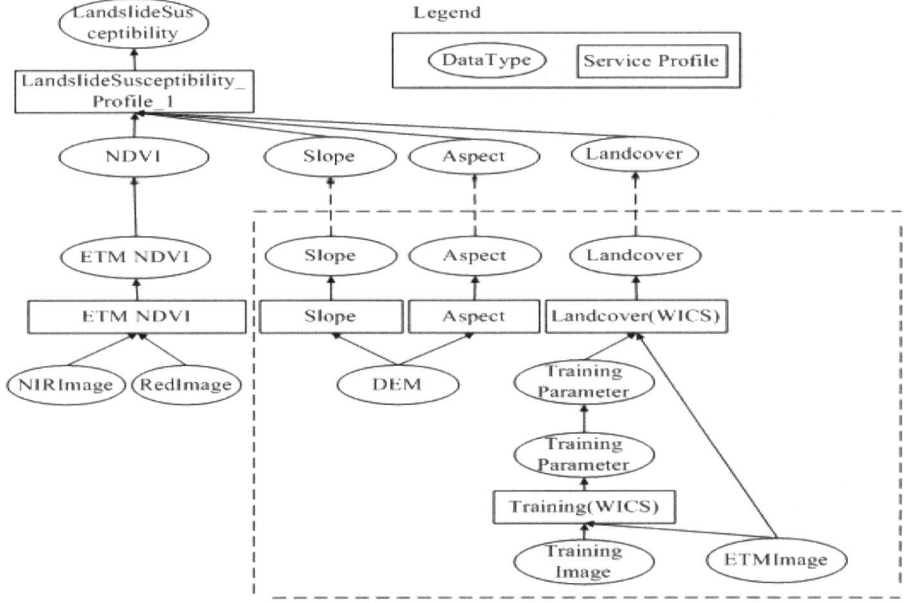

Fig. 3. An example of physical model

cannot be found, a matched service node in the graph can be selected to produce the requested data. Then the data query is moved on to the input "DataType" of the selected service node. The process continued until all input data are found available for the service chain. The resultant chain is called the "Physical Model". Figure 3 illustrates an example of the physical model resulting from the leaf node instantiation of an abstract model (i.e. logical path). The sub-chains inside the dotted rectangle represent the extension of the model after the instantiation process for a leaf node.

4.3.2 Service Instance Selection
Until now, each service node in the physical model is represented by an OWL-S service profile and is not bound to any service instance. A service profile can be bound to different service instances through corresponding service groundings. Different service instances are located at different physical addresses with related Quality of Service (QoS) information, such as network traffic and service performance. The selection of the service instance can be based on QoS information.

4.4 Service Chain Execution

The chaining result is represented as the OWL-S "Composite Process". It can be executed in an OWL-S engine. There are many XML-based service composition languages such as the Business Process Execution Language for Web Services (BPEL4WS), the Web Services Flow Language (WSFL) and the Web Service Choreography Interface (WSCI). Aalst [5] compared these common service composition languages from the aspect of control flow. Twenty flow control constructs, such as sequence, parallel split, and choice, were identified as the considerations most often required when designing a service composition language. The "Composite Process" ontology of OWL-S has control constructs from these pattern definitions. Composite processes are processes decomposable into other (non-composite or composite) processes. The decomposition is specified by using control constructs. Since the definitions of most control constructs originate from the service composition languages, a composite process can be converted to any of the service composition languages to enable execution in the existing engine for these languages.

5 Implementation and Result Analysis

5.1 Implementation

We use OWL-S to specify semantics for geospatial Web services. OWL-S API[5] is used for OWL-S parsing and grounding execution. Jena Transitive Reasoner[6] is selected for reasoning based on our application knowledge base. OWL-S Manager (OWLSManager), a component for OWL-S Files Management, is developed which can deploy and undeploy OWL-S files into the knowledge base. The path planning component (planner) operates on the service graph, which is dynamically generated from the run-time status of OWLSManager. The service profiles in OWL-S descriptions provide the nodes and edges in the graph as mentioned in section 4.1. Multiple paths as alternative plans are generated using the K-shortest path algorithm. A generation of Dijkstra's algorithm [6] is adopted. Since it is a label setting algorithms [6], paths are determined throughout the computations instead of at the ending of algorithm, thus the efficiency is ensured when a large number of services are involved. The individual services with the same service profile can be selected based on the QoS information. A unified interface for the QoS provider is supported so different QoS criteria can be plugged in. The instantiation process interacts with a grid-enabled CSW [7]. The chaining result, i.e., the OWL-S composite process, can either be executed directly or be converted to BPEL to facilitate the use of the BPEL engine. All these components work as Web applications. Figure 4 shows the user interface for path planning. An online demonstration of this implementation is available at http://www.laits.gmu.edu/geo/nga/index.html.

5.2 Result Analysis

We applied our implementation to the use case of landslide risk described in section 2 to test the effectiveness of this approach. The path modeling results from the source input "ETM Band 3" and the target output "Landslide Susceptibility" are shown in Figure 4 to answer the example question mentioned in section 4.2. Each logical path is visualized as a linked graph created by WebDot[7]. The applicapability of this method are demonstrated by automatically chaining of multiple Web services to derive the landslide susceptibility index of the certain area (Diamond Canyon, California) on a certain day. The SUBSUME match, other than the EXACT match, is

---

[5] http://www.mindswap.org/2004/owl-s/api/
[6] http://jena.sourceforge.net/inference/index.html
[7] http://www.graphviz.org/

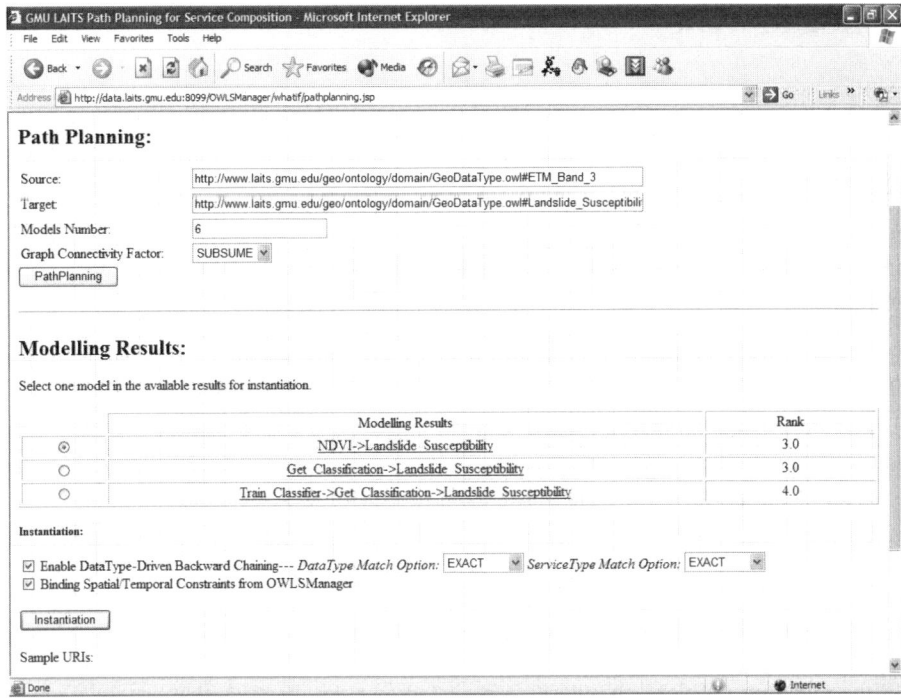

Fig. 4. The path planning user interface

capable to create available service chains, because the ETM NDVI service's output
ETM NDVI is not exactly the same as the NDVI input required by the landslide
susceptibility service. The service chain in this use case can be automatically and
dynamically generated whenever the CSW service is available and the data can be
queried using the CSW service. The composite process can also be registered in the
CSW as a virtual data product so that the composition process need not be repeated
when a new request for the same data product is submitted.

Two issues regarding this automatic path planning are identified through our
research and implementation experiment.

(1) Correctness of Model: The connection created in the service graph is purely based
on the matching of ontology descriptions of the input and output DataTypes without
considering the functionality of the services. It is possible that DataType ontologies
may not completely capture the underline differences in service functionality. Human
intervention might be needed to ensure the higher semantic accuracy for services in
the automatically generated models, e.g., users can decide which path is correct in the
"Modeling Results" part of Figure 4.

(2) Quality of Model: Currently, path selection is based on the length of the path. The
different match levels are reflected in the weight value of the connection edge, thus
making this criterion reasonable. However, the creditability of a service chain is also
affected by the internal implementation of services. Thus, a trust mechanism, such as
user rating system, can be introduced to help the selection of model.

6 Related Work

Ninja introduces the concept of a "path" to define an ordered sequence of services that, when composed, result in the desired complex service [8]. The path concept is further subdivided into logical path and physical path. A logical path is a sequence of compatible operators and connectors. Operators perform computations on data and connectors provide data transport between operators. A physical path is created when the operators in the logical path are implemented on a specific machine (service instance). XML-based input-output matching of services is used to generate a node and connection graph of the service chain, and BFS (Breadth-First-Search) algorithm is used to generate a logical path. A logical path is comparable to the process model in service composition, while the physical path is the instantiation of the process model. Our system uses a similar "path" concept. It differs in the following ways:

(1)   We focus on the geospatial domain and extend it with the data binding and instantiation process for leaf nodes.
(2)   We implement the K-shortest path to enable model selection.
(3)   We use a knowledge representation mechanism to describe the Web service and "DataType" match.
(4)   We use real Web services.

The problem of automatic service composition in the semantic Web areas through Artificial Intelligence (AI) planning has been discussed extensively [9] [10] [11] [12]. In the planning problem, the world is modeled as a set of states that can be divided into initial states and goal states. An action is an operation that changes one state to another state. Thus, the assumption for Web service composition as a planning problem is that Web service can be specified as an action with an initial state (preconditions) and a new state (effects). A common characteristic of these methods is that they are subject to constraints and assumptions that limit their use for wide applications. Special procedures need to be performed for each application. Our path planning approach captures this kind of geospatial Web service more appropriately for the data-centric geospatial problem, resulting in greater simplicity and efficiency.

Some instances of geospatial Web service composition have been reported [13], [14]. One example is the Geosciences Network (GEON) [13][15]. Geospatial Web services, including data (GML representation) provider services and customized services with vector data processing functionalities are sampled to compose a workflow manually in the KEPLER system [15]. The KEPLER system provides a framework for workflow support in the scientific disciplines. The major feature of the KEPLER system is that it provides high-level workflow design while at the same time hiding the underlying complexity of the technologies from the user as much as possible. Both Web service technologies and Grid technologies are wrapped as extensions in the system. For example, individual workflow components (e.g., data movement, database querying, job scheduling, remote execution) are abstracted into a set of generic, reusable tasks in the Grid environment [16]. Thus, combining a knowledge representation technique (e.g. OWL and OWL-S), with the lower level, generic, common scientific workflow tasks in the KEPLER system is a worthwhile technique for attempting to minimize or eliminate human intervention in the generation and instantiation of workflow. Thus, incorporation of our method in this system should be explored.

7 Conclusions and Future Work

This paper presents an approach to the automatic composition of geospatial Web services. OWL-S is adopted to describe a geospatial semantic Web service. Path planning is introduced to create service chain automatically. The service profiles in the OWL-S descriptions provide the nodes and edges in the graph. Multiple paths as alternative plans are generated using the K-shortest path algorithm. Individual services with the same service profile can be selected based QoS information. This system has been applied to a use case of landslide susceptibility assessment. Six types of Web services, including the OGC Web Coverage Service (WCS) and Web Image Classification Service (WICS), are involved in the final service chain. The study shows how this system is efficient and where it is applicable. Further study will focus on refinement of the plan to support more specified domain logics or rules and will also include addressing the potential improvements to model correctness and quality as described in the results analysis section.

Acknowledgements

This work is supported by a grant from U.S. National Geospatial-Intelligence Agency NURI program (HM1582-04-1-2021, PI: Dr. Liping Di).

References

1. Percivall, G., ed., 2002. The OpenGIS Abstact Specification, Topic 12: OpenGIS Service Architecture, Version 4.3, OGC 02-112. Open GIS Consortium Inc. 78 pp.
2. Gruber, T.R., 1993. A Translation Approach to Portable Ontology Specification, Knowledge Acquisition 5(2), pp. 199-220.
3. Dean, M., Schreiber, G., eds, 2004. OWL Web Ontology Language Reference, W3C. http://www.w3.org/TR/owl-ref.
4. Di, L., 2005. Customizable Virtual Geospatial Products at Web/Grid Service Environment. *Proceedings of International Geoscience and Remote Sensing Symposium, 2005 (IGARSS '05)*, IEEE International, Volume 6, pp. 4215 – 4218.
5. Aalst, W., 2003. Don't Go with the Flow: Web Services Composition Standards Exposed, IEEE intelligent systems, January/February (2003), pp. 72-76.
6. Martins, E.Q.V., Pascoal, M.M.B. and Santos, J.L.E., The K Shortest Paths Problem, Research Report, CISUC, June 1998.
7. Wei Y., Di L, Zhao B., Liao G., Chen A., Bai Y., Liu Y., 2005. The Design and Implementation of a Grid-enabled Catalogue Service. *Proceedings of International Geoscience and Remote Sensing Symposium, 2005 (IGARSS '05)*, IEEE International, Volume 6, pp. 4224 – 4227.
8. Chandrasekaran, S., Madden, S., and Ionescu, M., 2000. Ninja Paths: An Architecture for Composing Services over Wide Area Networks, CS262 class project writeup, UC Berkeley, 2000. 15 pp.
9. McIlraith, S. A., Son, T. C., 2002. Adapting Golog for Composition of Semantic Web Services. In D. Fensel, F. Giunchiglia, D. McGuinness, and M.-A. Williams, editors, Proc. of the 8th International Conference on Principles and Knowledge Representation and Reasoning (KR'02), France, 2002. Morgan Kaufmann Publishers. pp. 482—496.

10. Wu, D., Parsia, B., Sirin, E., Hendler, J., Nau, D., 2003. Automating DAML-S web services composition using SHOP2. In Proceedings of 2nd International Semantic Web Conference (ISWC2003), Sanibel Island, Florida, October 2003. 16 pp.
11. Klusch, M., Gerber, A., Schmidt, M., Semantic Web Service Composition Planning with OWLS-Xplan, Agents and the Semantic Web,2005 AAAI Fall Symposium Series, Arlington, Virginia, USA, 4th - 6th November, 2005. 8 pp.
12. Di, L., 2005. A Framework for Developing Web-Service-Based Intelligent Geospatial Knowledge Systems, *Journal of Geographic Information Sciences.* Vol. 11, no 1, pp. 24-28.
13. Jaeger, E., Altintas, I., Zhang, J., Ludäscher, B., Pennington, D., Michener, W., 2005. A Scientific Workflow Approach to Distributed Geospatial Data Processing using Web Services, 17th International Conference on Scientific and Statistical Database Management (SSDBM'05), 27-29 June 2005, Santa Barbara, California. pp. 87-90.
14. Di, L., Zhao, P., Yang, W., Yu, G., Yue, P., 2005. Intelligent geospatial web services *Proceedings of International Geoscience and Remote Sensing Symposium, 2005 (IGARSS '05)*, IEEE International, Volume 2, pp. 1229 - 1232
15. Ludäscher, B., Altintas, I., Berkley, C., Higgins, D., Jaeger, E., Jones, M., Lee, E., Tao, J., Zhao, Y., 2005. Scientific Workflow Management and the Kepler System, *Concurrency and Computation: Practice & Experience*, Special Issue on Scientific Workflows, to appear, 2005. 19pp.
16. Altintas, I., Birnbaum, A., Baldridge, K., Sudholt, W., Miller, M., Amoreira, C., Potier, Y., Ludäscher, B., 2004. A Framework for the Design and Reuse of Grid Workflows, Intl. Workshop on Scientific Applications on Grid Computing (SAG'04), Herrero, P., Prez, M.S., and Robles, V. (Eds.): SAG 2004, LNCS 3458, Springer, pp. 119 – 132.

Indexing Moving Objects on Road Networks in P2P and Broadcasting Environments

Hye-Young Kang, Jung-Soo Kim, and Ki-Joune Li

Department of Computer Science and Engineering, Pusan National University,
Pusan 609-735, South Korea
{hykang, jskim}@isel.cs.pusan.ac.kr, lik@pnu.edu

Abstract. Scalability is one of the crucial problems in realizing massively distributed systems such as ubiquitous computing. In this paper, we focus on indexing methods in massively distributed environments. A number of work on indexing in P2P, like CAN and Chord, have been devoted to overcome this problem. The lengths of routing path are $O(dn^{\frac{1}{d}})$ for CAN and $O(\log n)$ for Chord, which are in fact the cost of search, where there are n nodes. In this paper, we propose an alternative indexing scheme not only relying on P2P but also on broadcasting environments. The contributions of this paper include firstly the reduction of routing path to nearly $O(1)$ for road-oriented query by using broadcasting, and secondly handling the mobility of nodes on road networks.

1 Introduction

Peer-to-Peer (P2P) is a promising approach to overcome the scalability problem of massively distributed systems and ubiquitous computing environments. Several methods have been proposed based on P2P such as Chord [1] and CAN [2]. They provide efficient searching mechanism via distributed lookup tables, which are effectively distributed indices. But they have several weak points in applying it to applications related to GIS or spatial database systems. First, they are not capable of handling mobility of nodes, even though some of them provides geographic routing and searching [3,4]. Only static locations of nodes are considered by these methods. Second, the length of routing path and hop counts of these methods, which are basically the cost of searching in P2P, are still large.

In this paper, we propose a distributed indexing methods for mobile nodes in P2P environments, where a fraction of index is periodically broadcasted over the mobile nodes. This hybrid approach of indexing with P2P and broadcasting reduces the hop counts nearly to a constant. But due to the limited bandwidth of broadcasting, the index structure to be broadcasted should be designed carefully with minimum size of data as possible. The large size of broadcasting data influences on the performance of search as indicated in [5]. For this reason, we only contain a small fraction of indices in broadcasting messages.

In real applications, most mobile nodes are found on road networks. For example, vehicles are typical example of mobile nodes and they move on road networks.

J.D. Carswell and T. Tezuka (Eds.): W2GIS 2006, LNCS 4295, pp. 227–236, 2006.

However the previous studies on mobile have focused on euclidian space except [6,7]. In this paper, we rather focus on mobile objects in road network spaces.

The results of experiments with a real road network data set show that the performance of our method rely on the speed of mobile objects and the frequency of broadcasting and most in-network queries can be processed within four hop counts.

The rest of this paper is organized as follows; in the next section, we present related work and motivations of our research. We present the basic data structures mechanisms of our indexing method in section 3. The algorithms for handling the mobility of nodes and query processing are given in section 4. In section 5, we analyze the performance of our method with a cost model and experiments with a real road network data and simulator. In section 6, we conclude the paper.

2   Related Work and Motivations

In this paper, we assume that each node has an IP address so that it can directly send a message to others if it knows their IP addresses. Based on this assumption, a number of methods have been proposed to process search queries based on distributed hash tables (DHT) such as Tapestry [8], Chord [1], and CAN [2]. By these methods, each node has a small fraction of distributed lookup tables to index. They were initially intended for processing exact match queries like keyword search. Several extensions have been made to process range query as well. For example, an extension of CAN has been tried to handle range queries [9]. It maps multi-dimensional range to an 1-D key by using Hilbert curve. By similar way, MAAN also extends Chord to process multi-dimensional range queries by introducing a uniform locality preserving hashing function [10]. Even though they are capable of processing spatial query, they have several weak points.

First, they are not intended to process spatial queries, since they do not fully consider spatial properties, such as spatial proximity. For this reason, their processing cost, which is mainly measured by routing hop counts, is far from being optimal. For example, the processing cost of range query for MAAN is $O(\sum_{i=1}^{M}(\log N + s_i N))$, where M is the number of attributes, N is the number of nodes, and s_i is the selectivity for the i-th attribute. $M = 2$, since we are dealing with spatial query [10]. It implies that the cost is approximately a double of the processing cost for 1-dimensional range query. PePeR is also an indexing method for P2P and supports range query [11]. On the other hand, the theoretical lower bound of routing hop counts of PePeR is known as $\frac{4}{d}\sqrt[d]{N}$, where d means the number of dimensions and N is the number of nodes. It grows relatively rapidly when N becomes large. In fact, the ideal cost for processing spatial query should be constant regardless of the number of nodes.

Second, they do not fully support the mobility of nodes, since the update cost of attributes and locations is relatively expensive. Furthermore, the movement of mobile nodes is not taken into account by these methods. For example, it is no longer valid to process spatial query for mobile objects via finger tables of Chord, since the values of finger table computed from the positions of mobile nodes at a time instance are no longer at the positions.

Third, these methods are intended to process the spatial query in euclidian space, while the mobile object of most applications are on road network space. A number of properties of road network space differ from the euclidian space, such as the definition of distance, the linear reference systems, etc. For example, DisTIN [12], which is a distributed indexing method for P2P environment and supports the mobility, is limited to euclidian space and not applicable to road network space.

For these reasons, our work aims to develop an indexing method for mobile objects on road networks with the following considerations;

- to reduce the routing hop counts to $O(1)$, and
- to support the mobility on road networks.

In order to achieve the above objectives, we apply digital broadcasting environments, which have been already in service such as DvB (Digital Video Broadcasting), or DMB (Digital Multimedia Broadcasting).

3   Distributed Spatial Indexing Using P2P and Broadcasting: DIMOR-PnB

The performance of spatial indexing in P2P environments is mainly determined by routing hop counts from the node where the query is issued to the node containing the information about the answers. The routing hop counts increase according to the distance between the query and answer nodes. The basic idea of our method will be explained in this section, which provides a routing mechanism with a constant hop count via broadcasting. Our method is based on the following assumptions.

- All nodes move road networks.
- Each node has its IP address.
- Each node can directly communicate with others by IP address.
- Each node has the information on its location, IP address, and optionally the locations and IP addresses of its neighbor nodes
- A broadcasting server collects the information of a set of selected nodes and periodically broadcasts them.
- A query region is specified by road segment.

Figure 1 gives a brief sketch on the idea of our method, called DIMOR-PnB(Distributed Indexing Moving Objects on Road network using P2P and Broadcasting). Suppose that a query "*Find the nodes within a distance* d *from a point* Q" is given to a node P_1. If P_1 has only the information of P_2 like figure 1-a, there is no other choice except forwarding the query message as $P_1 \rightarrow P_2 \rightarrow P_7 \rightarrow P_8 \rightarrow P_9$. On the contrary, if a broadcasting server broadcasts the information of mobile nodes like figure 1-b, P_1 can choose P_{10} as a short cut of routing the query message. Then the message is forwarded as $P_1 \rightarrow P_{10} \rightarrow P_9 \rightarrow P_8$.

The data structures of DIMOR-PnB consist of two parts; firstly broadcasting messages and secondly a index stored at each super node. They will be explained in the next subsections.

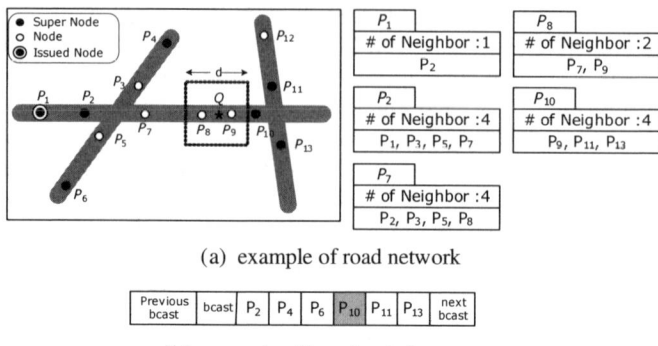

(a) example of road network

| Previous bcast | bcast | P_2 | P_4 | P_6 | P_{10} | P_{11} | P_{13} | next bcast |

(b) example of broadcasted message

Fig. 1. Example of spatial query processing

3.1 Selecting Super Nodes

The ideal approach would be to let the server broadcast the information of all nodes, so that each node could get the location and IP address of any node and process spatial query by only one hop. In practice, it is impossible since the bandwidth of broadcasting is limited and only a small fraction of band can be allocated to a specific service. Furthermore, the increase of data to broadcast may increase the period of broadcasting and result in an inaccuracy of dynamic information on mobile nodes.

In stead of broadcasting the information of all nodes, we select a certain number of super nodes from the nodes, where the server broadcasts only the information of these super nodes. In this paper, we apply the following rules to select super nodes.

- A super node is to be selected per a road segment.
- If a node enters into a road segment where there is no node, then it becomes the super node.
- If the super node on a road segment moves to other segment, the node that has entered to the node least recently becomes the succeeding super node.

The number of super nodes to be selected is slightly less the number of road segments. We can reduce the number of super nodes by merging unnecessarily segmented roads. Suppose that the data for one super node consists of its IPv6 address and location, then the size becomes 14 bytes per super node. For instance, a metropolitan like Pusan consists of approximately 20,000 road segments. It means that the size of broadcasting message for DIMOR-PnB is about 280 K bytes, which is a reasonable size for terrestrial DMB.

3.2 Management of Local Topology by Neighbor Table

Since the broadcasting message does cover only a fraction of index, a complementary index is required. For this reason, each super node keeps a table, called *neighbor table* consisting of the locations and IP addresses of the nodes on its road segment. The data structure of neighbor table is shown in figure 2. Note that the location on a road segment is specified by linear reference systems like *offset* from the start point rather than (x, y)-coordinate systems. The last two fields, `# of successors` and `(IP, Road ID)*` of the neighbor table will be explained in section 4.

Node
IP,(Road ID,offset)
# of Neighbor
(IP,offset)*
# of successor
(IP, Road ID)*

Fig. 2. Data structure of neighbor table

Due to the mobility of nodes, we need to maintain their dynamic membership as well as their location, and the algorithms are presented as follows.

Node Entering: When a node enters to new road segment, the node must listen the next broadcasting message to check existence of super node. In the first case where no super node is found on the new road segment, the node becomes super node and registers to server. Figure 3 shows an example of movement of nodes on road networks. When a node P_2 move to $S1$ from $S4$ at t_1, P_2 checks existence of super node on $S1$. Since there is no node on $S1$, P_2 registers itself to the server by sending its location and IP address. Then the broadcasting message will be changed at t_2 as shown by figure 3-b. Second, in case that there is already a super node on the road segment, the node sends its location and IP address to the super node. For example, when P_7 move to $S3$ from $S4$ at t_2, P_7 checks the broadcasting message and gets the information of the super node P_4. Then, P_7 sends its information to P_4.

Node Leaving: When a node moves to other road segment or disappears, the node should inform its leaving to the super node. In the first case where the leaving node is not the super node, the node should notify its leaving to the super node. For example, when P_2 leaves $S4$ at t_1 in figure 3-a, P_2 is removed from the neighbor table of super node P_5 as shown by figure 4-a and 4-b. In the second case where the leaving node is a super node, it has to select its successor among the neighbor nodes and transfer the neighbor table to the successor. The change of the finger table for super node P_5 is shown by figure 4-c and figure 4-d. The change of super node must be informed to the broadcasting server as well. If there is no neighbor node on the road segment, the leaving node just asks the server to delete itself from the broadcasting message.

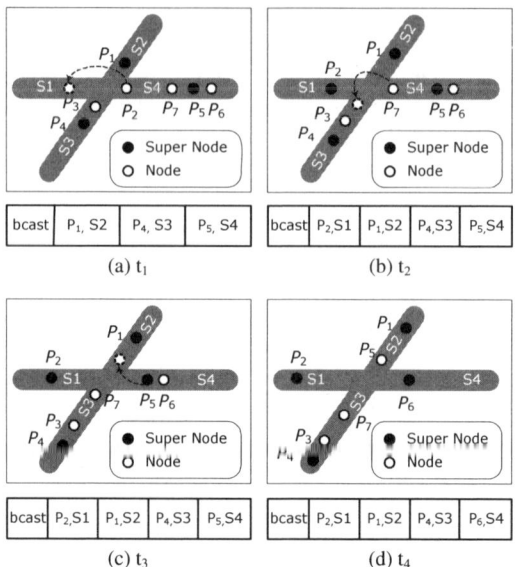

Fig. 3. Example of movement of nodes on the road networks

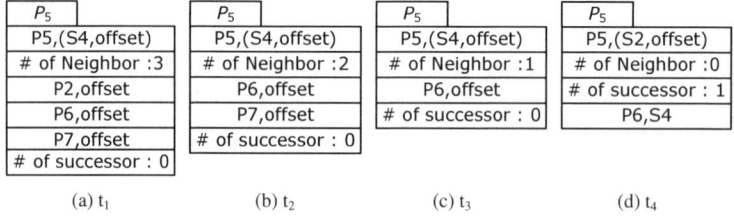

Fig. 4. Change of the neighbor table of P_5

Movement on the same road segment: When a node moves on the same road segment, its position of nodes continuously changes, and should be periodically reported to the super node for correct query processing. For this, each node reports own location to super node.

4   Spatial Query Processing Based on DIMOR-PnB

In this section, we will present how to apply DIMOR-PnB for processing spatial query by an example of nearest neighbor(NN) query. Other types of spatial queries can be processed by similar way. Suppose that a query "*Find the nearest node to a given point q on road networks.*" is submitted to a node P_1 as shown by figure 5-a. Then this query is to be processed by four steps with DIMOR-PnB as follows.

- **step 1:** P_1 searches the super node that is on the same road segment of the query point q from the recently received broadcasting message.
- **step 2:** P_1 sends the query message to the super node P_7 founded by step 1.
- **step 3:** P_7 searches the nearest node to q
- **step 4:** P_7 sends the location and IP address of P_9, the nearest node to q, to the P_1.

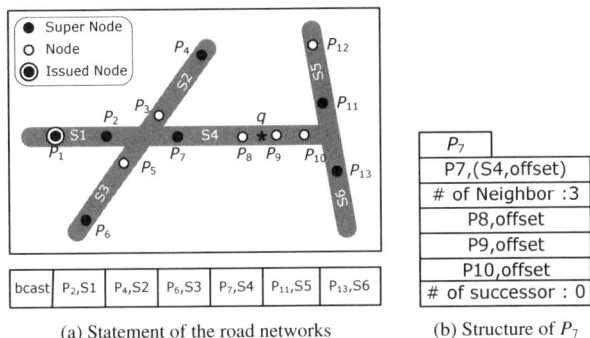

(a) Statement of the road networks   (b) Structure of P_7

Fig. 5. Example of Nearest Neighbor Query Processing

In some cases where the speed of node is high, there may be a mismatch between the actual position of a super node and its location data in the recent broadcasting message. It may happen when a super node leaves from a road segment after having reported its location to the server. Then location data on the broadcasting message may be no longer valid. In order to avoid this problem, the leaving super node should keep the pointer to its succeeding super node as depicted by figure 2. Note that the successors may be more than one, when the speed is very high. Figure 6 illustrates this case of query processing, where the super node on the road segment has been changed.

5 Experiments

In order to analyze the performance of our method, we performed several experiments with a real data set of the road networks in Pusan and three synthetic data sets of 100,000 mobile nodes generated with different average speeds, 40Km/h, 60Km/h, and 80Km/h respectively by [13]. We executed 500 nearest neighbor queries to evaluate the performance of our method. We considered routing hop counts as the cost measure of DIMOR-PnB.

Figure 7-a shows the costs of the nearest neighbor query processing. The lower bound of routing hop counts for DIMOR-PnB is 2, since we need one hop for step 2 in section 4 and one more hop for returning answer of query. The additional costs are mainly caused by leaving super nodes. If it is no longer on the same road segment, the actual super node should be retrieved by traversing the successor pointers of neighbor tables. Although the additional cost increases according to the increase of speed, the results show that the increases are very gradual.

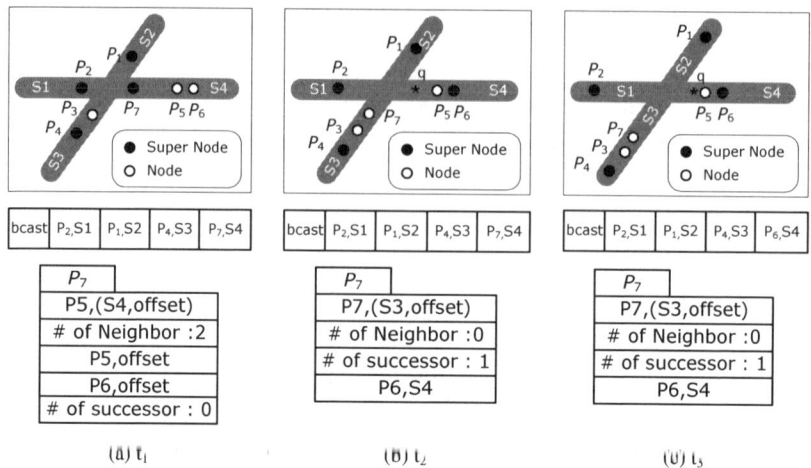

Fig. 6. Example of nearest neighbor query processing via successors

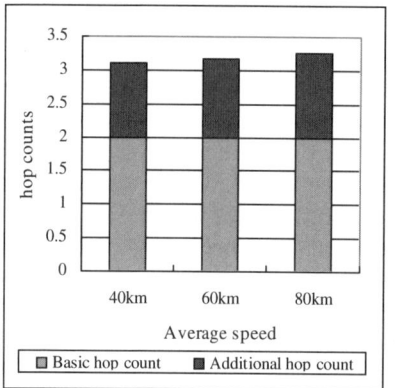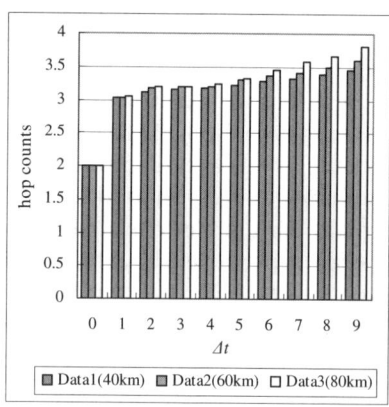

(a) NNQ cost with respect to average speed

(b) NNQ cost with respect to $\triangle t$

Fig. 7. Query processing costs

Figure 7-b shows the cost according to the differences between the times of query and broadcasting. When there is no difference, it means that the location data in broadcasting message is correct and no additional cost is required regardless of the speed. As the difference increases, the cost increases and the cost differences between different speeds become apparent. However the increase according to the time difference is extremely slow. We observe a big jump of additional cost between $\triangle t = 0$ and $\triangle t = 1$, which is due to leaving super nodes. It means that we will observe another big jump when the leaving super node will change one more road segment.

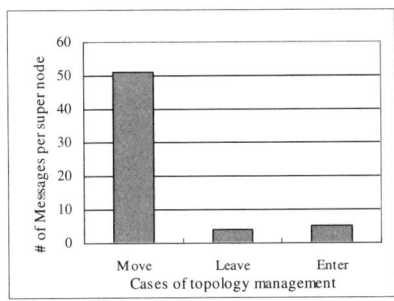

Fig. 8. Number of messages per super node

Most of messages that a super node receives are for updating locations of nodes without changing road segment, as shown by figure 8. It is expected that we could reduce the number of messages by using more accurate tracking methods such as [14].

6   Conclusion

Although a number of methods have been proposed to process spatial queries, very few attentions have been paid on spatial query process based on road networks in P2P environment.

In this paper, we propose a method for spatial indexing and processing spatial query by a hybrid approach composed of P2P and broadcasting. The contributions of our work are the development of a distributed spatial indexing method by the hybrid approach 1) for mobile nodes, 2) in road network space, and 3) the reduction of routing hop counts to almost constant in comparison with the previous P2P indexing method.

Acknowledgment. This work was supported by the Korea Research Foundation Grant Funded by the Korea Government(MOEHRD)(KRF-2006-209-D00008), by IRC(Internet Information Retrieval Research Center) in Hankuk Aviation University. IRC is a Regional Research Center of Kyounggi Province, designated by ITEP and Ministry of Commerce, Industry and Energy.

References

1. Ion Stoica, Robert Morris, David R. Karger, M. Frans Kaashoek, and Hari Balakrishnan. Chord: A scalable peer-to-peer lookup service for internet applications. In *Proceedings of International Conference on Applications, Technologies, Architectures, and Protocols for Computer Communication*, pages 149–160, 2001.
2. Sylvia Ratnasamy, Paul Francis, Mark Handley, Richard M. Karp, and Scott Shenker. A scalable content-addressable network. In *Proceedings of International Conference on Applications, Technologies, Architectures, and Protocols for Computer Communication*, pages 161–172, 2001.

3. Brad Karp and H. T. Kung. GPSR: Greedy Perimeter Stateless Routing for wireless networks. In *Proceedings of International Conference on Mobile Computing and Networking*, pages 243–254, 2000.
4. Young-Bae Ko and Nitin H. Vaidya. Location-Aided Routing (LAR) in mobile ad hoc networks. *Wireless Networks*, 6(4):307–321, 2000.
5. Tomasz Imielinski, S. Viswanathan, and B. R. Badrinath. Data on Air: Organization and Access. *IEEE Transactions on Knowledge and Data Engineering*, 9(3):353–372, 1997.
6. Christian S. Jensen, Jan Kolárvr, Torben Bach Pedersen, and Igor Timko. Nearest neighbor queries in road networks. In *Proceedings of ACM International Symposium on Advances in Geographic Information Systems*, pages 1–8, 2003.
7. Christian S. Jensen, Torben Bach Pedersen, Laurynas Speicys, and Igor Timko. Data Modeling for Mobile Services in the Real World. In *Proceedings of International Symposium on Spatial and Temporal Databases*, pages 1–9, 2003.
8. Ben Y. Zhao, John Kubiatowicz, and Anthony D. Joseph. Tapestry: a fault-tolerant wide-area application infrastructure. *Computer Communication Review*, 32(1):81, 2002.
9. Artur Andrzejak and Zhichen Xu. Scalable, Efficient Range Queries for Grid Information Services. In *Peer-to-Peer Computing*, pages 33–40, 2002.
10. Min Cai, Martin R. Frank, Jinbo Chen, and Pedro A. Szekely. MAAN: A Multi-Attribute Addressable Network for Grid Information services. In *Proceedings of International Workshop on Grid Computing*, pages 184–191, 2003.
11. Antonios Daskos, Shahram Ghandeharizadeh, and Xinghua An. PePeR: A Distributed Range Addressing Space for Peer-to-Peer systems. In *Proceedings of International Workshop on Databases, Information Systems, and Peer-to-Peer Computing*, pages 200–218, 2003.
12. Hye-Young Kang, Bog-Ja Lim, and Ki-Joune Li. P2P Spatial Query Processing by Delaunay Triangulation. In *Proceedings of International Workshop on Web and Wireless Geographical Information Systems*, pages 136–150, 2004.
13. Bo-Ryun Kim. Generation of the Vehicle Trajectories using the Realistic Speed Model. Master's thesis, Pusan National University, 2006.
14. Alminas Civilis, Christian S. Jensen, Jovita Nenortaite, and Stardas Pakalnis. Efficient Tracking of Moving Objects with Precision Guarantees. In *Proceedings of International Conference on Mobile and Ubiquitous Systems*, pages 164–173, 2004.

GeoComputation in the Grid Computing Age

Qingfeng Guan[1], Tong Zhang[2], and Keith C. Clarke[1]

[1] Department of Geography, University of California, Santa Barbara
Santa Barbara, CA 93106, USA
{guan, kclarke}@geog.ucsb.edu
[2] Department of Geography, San Diego State University
5500 Campanile Drive, San Diego, CA 92182-4493, USA
zhangt@rohan.sdsu.edu

Abstract. This paper first discusses some challenges that current GeoComputation faces in terms of usability, feasibility, applicability and availability, and the opportunities that will arise when new computing technologies, especially Grid Computing, emerge and prevail. A Grid-based geospatial problem-solving architecture is proposed to provide a solution for building an easy-to-use, widely accessible and high-performance geospatial problem-solving environment that integrates multiple complicated GeoComputational processes at an acceptable cost. A parallel geographic cellular automata model is given as an example to address some distinguishing issues when designing and implementing parallel algorithms for GeoComputation to effectively and efficiently utilize the computational Grid.

Keywords: GeoComputation, Grid Computing, Web portals, Parallel algorithm design.

1 Introduction

GeoComputation, used as a term since the first GeoComputation conference in 1996, has been defined as the "*art and science of solving complex spatial problems with computers*" (www.geocomputation.org). After a decade of study and development, GeoComputation has drawn a great deal of attention from not only geographers but also researchers and professionals in other disciplines. A wide range of computational techniques and methods have been introduced and deployed to solve complex geospatial problems and a large number of applications and demonstrations have been presented in the GeoComputation conferences and publications.

In terms of problem domains and research interests, GeoComputation has shown high diversity, which on one hand reflects GeoComputation's increasing institutional recognition and interests. On the other hand, this diversity makes GeoComputation appear as a "*grab-bag*" of computational techniques for geospatial problem-solving [1] lacking in interoperability, and extremely hard to organize to form an integrated geospatial problem-solving environment.

Most of the computational techniques and methods used in GeoComputation introduce high computational loads. Increasingly complicated geospatial models

J.D. Carswell and T. Tezuka (Eds.): W2GIS 2006, LNCS 4295, pp. 237–246, 2006.

usually involve multiple GeoComputational processes to simulate complex geospatial phenomena. Also, the ever-growing amount of high-resolution geospatial and detailed non-geospatial data from a wide range of devices and agencies, coupled with the increasing requirements for accuracy, make computational performance an inevitable concern in geospatial problem solving.

Recent emerging computing technologies, especially Grid Computing, provide solutions to tackle these challenges and are promising to promote GeoComputation in terms of usability, feasibility, applicability and availability. Grid Computing, a new network-based computing technology, is concerned with *"direct access to computers, software, data, and other resources as is required by a range of collaborative problem-solving and resource brokering strategies"* [2]. It is widely believed that Grid-based web services and web portals have the prospects of improving the interoperability of geospatial data and operations and providing easy-to-use interfaces for end users. Grid-enabled parallel computing can be used to provide widely accessible high-performance computational capacity. However, to deploy these Grid-based computing technologies, i.e., web services, web portals and parallel computing, for GeoComputation, some distinguishing characteristics of geospatial data and Geoprocessing have to be considered and dealt with carefully.

This paper firstly discusses the challenges and opportunities that current GeoComputation faces in the Grid Computing context. After a brief summary of recent progress in Grid-based web services, web portals and Grid-enabled parallel computing, a Grid-based geospatial problem-solving architecture is proposed as a comprehensive solution for Grid-enabled GeoComputation, and a parallelized geographic cellular automata (Geo-CA) model is presented as an example to discuss the concerns and approaches during the design and implement of a Grid-compatible parallel GeoComputation application.

2   GeoComputation: Challenges and Opportunities

2.1   The Capital G and C in GeoComputation

Couclelis [1] identified the "core GeoComputation" as innovative (or derived from other disciplines) computer-based geospatial modeling and analysis, contrasted against traditional computer-supported spatial data analysis and geospatial modeling including routine GIS research and applications. GeoComputation's distinction lies in the underlying computational emphases. Couclelis [1] argued that GeoComputation should be considered based on the theory of computation, which is mainly concerned with effective procedures. Openshaw [3] also emphasized the computational science as the origin of GeoComputation (the Computation part) and the essential concerns about geographical and earth systems (the Geo part). In this sense, GeoComputation is *"a form of computational science applied to spatial or geo-problems (theory and data) … G[eo]C[omputation] is not just an add-on to GIS … is concerned with the applications of a computational science paradigm to study all manner of geo-phenomena including both physical and human systems"* [3].

The capital G and C, as the core of GeoComputation, fundamentally explain the philosophical origins of GeoComputation, which is the revolutionary application of computational science in geography or an even broader domain.

2.2 Challenges and Opportunities

Couclelis [1] identified five major challenges that proponents of GeoComputation have to tackle: 1. To develop major demonstration projects of obvious practical interest, and to attract commercial companies to invest in and market products of GeoComputation; 2. To develop scientific standards compatible with mainstream quantitative geography and other Geosciences; 3. To constitute epistemological definition of GeoComputation to make it to *"appear like little more than a grab-bag of problem-solving techniques of varying degrees of practical utility"*; 4. To justify the 'geo' prefix by developing a coherent perspective on geographical space;   5. To move GeoComputation in the directions which parallel *"the most intellectually and socially exciting computational developments"*.

The content of the recent GeoComputation conferences demonstrates the high diversity of the research in GeoComputation. This, on one hand, reflects GeoComputation's increasing recognition in geography and other disciplines. On the other hand, such diversity makes GeoComputation appear, still, as a "grab-bag" of computational techniques for geospatial problem solving. Even after a decade of studies and development, the challenges discussed by Couclelis still remain.

Despite the epistemological and theoretical aspects of her discussion, let us take a look at these challenges from a computing perspective:

1. Currently, most GeoComputation research and applications focus on applying a specific computational technique or method in a specific problem domain. However, to analyze, simulate and even forecast complex geospatial phenomena, geospatial models usually need to utilize multiple complicated GeoComputational processes. However, the high diversity of GeoComputation and the complexities of geospatial models make it hard to effectively and efficiently organize a wide range of GeoComputational processes to form an easy-to-use integrated geospatial problem-solving environment to provide solutions for real-world applications.

2. Many of the computational techniques and methods used in current GeoComputation are derived from *"the field of artificial intelligence (AI) and the more recently defined area of computational intelligence (CI) … include expert systems, cellular automata, neural networks, fuzzy sets, genetic algorithms, fractal modelling, visualization and multimedia, exploratory data analysis and data mining"* [1]. As discussed above, the underlying innovation of GeoComputation is to apply computational science, which focuses on effective procedures, in the geospatial context. However, most of these techniques and methods are computationally intensive when it comes to real-world applications due to the extreme complexities of geospatial problems. Furthermore, the wide adoption of new technologies for geospatial data measuring and collecting has been bringing us a huge amount of geospatial data. The prevalence of cross-discipline research activities and ever-increasing requirements for accuracy have been pushing and improving the development of more sophisticated and complicated geospatial

models which again require intensive computation. On the other hand, the high costs of high-performance supercomputers keep them as high-end facilities that are hardly available and accessible for GeoComputation practitioners.

Fortunately, new computing technologies provide a possible solution to tackle these challenges. Grid Computing, an emerging network-based computing technology, has shown a great prospect for building a widely accessible platform (or cyberinfrastructure) for seamless integration of computing resources across the network. Research and development of web services and web portals based on the Grid are promising to improve the interoperability of data and operations and to provide easy-to-use interfaces. By connecting computing resources on the networks, Grid Computing also provides high-performance computational capacity at a much lower cost. In one word, Grid Computing offers us an opportunity to promote GeoComputation in terms of usability, feasibility, applicability and availability. In spite of the great deal of passion and interest that have been put into Web GIServices, Semantic GIS and Grid GIS, research on Grid-enabled GeoComputation is largely lacking. A very few research activities and applications exist currently. GIsolve at the University of Iowa (http://grow.its.uiowa.edu/) can be seen as one. GIsolve, as a Grid-enable geospatial problem-solving environment, supports spatial interpolation, statistics and domain decomposition.

3 Emerging Technologies to Promote GeoComputation

3.1 Grid Computing

Grid technologies have been rapidly evolving since the mid1990s. They are primarily concerned with the issues on the integration of large-scale computational resources and services [4]. The increasing diversity of computational and human resources created the "Grid problem" which requires dynamical resource sharing mechanism between "Virtual Organizations" [2]. "Virtual Organization" is in particular a ground-breaking concept for crossing the administrative and institutional boundaries for resource and services sharing. This also is of significance for cyberinfrastructure construction by forming network-based dynamic communities. In addition to decentralized control, common and universal protocols plus "quality services" are the two characteristics that distinguish the Grid from other technologies such as cluster computing, peer-to-peer computing, distributed computing [2, 5].

3.2 Web Portals for the Grid

People gradually realized that to make Grid technologies more widely used by the scientific community, it is critical to simplify the low-level configurations. The web portal offers a centralized and uniform interface to access the distributed and heterogeneous resources and services. Web portals have been widely implemented in many commercial websites to offer personalized web-based services such as personalized web news, web calendars et al. Web portals facilitate the use of these services by saving time that previously had to be spent on finding the separated services. Scientists can borrow the idea and use the web portal as an easy-to-use interface between end users and Grid systems/resources.

3.3 Grid-Enabled Parallel Computing

A computational Grid is "*a hardware and software infrastructure that provides dependable, consistent, pervasive, and inexpensive access to high-end computational capabilities*" [6]. Flexibly defined assemblages of computing resources (supercomputers, workstations, PCs, etc.) connected by high-performance networks (LAN, WAN, Internet, etc.) are used to form a distributed parallel computing environment. These assemblages are supposed to be transparent to users. That means, just like using the electricity grid, users use the computational Grid as a "virtual supercomputer", and do not need to be aware of where or what the computing resources are.

Globus Toolkit, developed by the Globus project (www.globus.org), has become the *de facto* standard for Grid-based scientific research and industrial products. TeraGrid (www.teragrid.org), a prototype computational Grid, is the world's largest, fastest, distributed infrastructure for open scientific research.

4 Strategies and Approaches

4.1 A Grid-Based Geospatial Problem-Solving Architecture

To organize multiple complicated GeoComputational processes and to form an integrated and easy-to-use environment, a Grid-based four-tier geospatial problem-solving architecture is proposed, which is shown in Fig. 1.

The first tier is the *Presentation* tier which is the interface between users and the underlying distributed computing environment. Typically, a GIS/GeoComputation web portal can provide visualization tools for data searching, spatial analysis, and the display of final GeoComputation results. The web portal can also become a Visual Problem-Solving Environment (VPSE) in which users can visually and interactively manipulate active online GeoComputation services as analysis modules. That is to say, users can implement their personalized analysis strategies by using a GIService composer to assemble heterogeneous sources on the Grid. The GIService composer displays job monitoring information as well as the resource scheduling information for every individual GIService.

The *Service* tier, which connects the *Presentation* and the *Model* tier, provides a mechanism to locate and fetch requested GIServices. These services can be hosted by anyone across the Grid, but have to be OGC-compatible web services and Grid-enabled. Automatically collecting and locating the most desired services are critical. Highly intelligent and efficient GIS filters should be developed to reduce the burden of GIService locators to find the matching services in accordance with the requests from end users.

In the *Model* tier, Grid-enabled parallel GeoComputation algorithms and models can be published as GIServices in the *Service* tier. GeoComputation researchers should take into account the special needs and characteristics of GeoComputation to develop

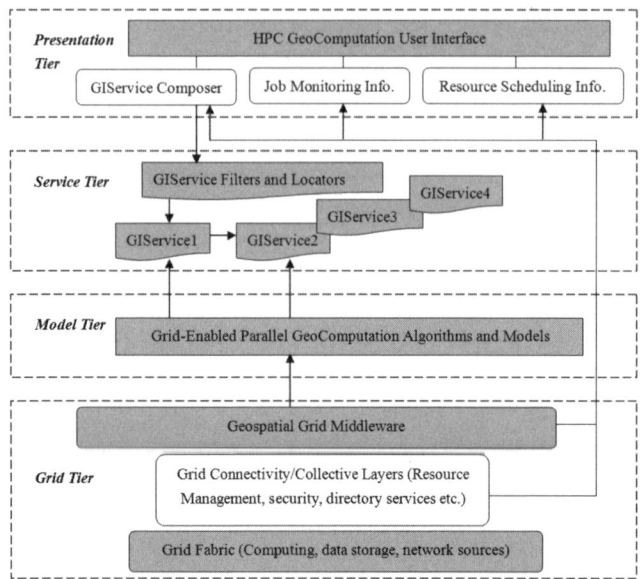

Fig. 1. A Grid-based GeoComputation web portal problem-solving architecture

highly efficient parallel algorithms and models. A parallel GeoComputation design tool should be provided to allow users to develop Grid-compatible parallel algorithms and models.

The last tier, the *Grid* tier, functions as the underlying infrastructure to support the high-level GIS/GeoComputation services. This infrastructure includes all the underlying Grid Computing hardware and software. They provide a supportive environment to make high-performance GeoComputation possible. Above the Grid infrastructure, a geospatial Grid middleware should be developed to extend the Open Grid Services Architecture (OGSA) with the geospatial web service specifications.

4.2   Case Study: Grid-Enabled Geographic Cellular Automata

Geographic Cellular Automata (Geo-CA) are typical GeoComputation models, and have been widely used for simulating and forecasting complex spatial-temporal phenomena. For most Geo-CA models, calibration processes are needed to determine the appropriate values of a set of parameters, which indicate multiple geospatial and non-geospatial factors, so that the models can produce realistic simulation results. However, due to the very large number of combinations of parameter values and the massive amount of high-resolution geospatial data, this calibration process is usually extremely time-consuming [7]. This section discusses some issues and approaches during designing and implementing a parallel Geo-CA model in a computational Grid context, i.e., the proposed Grid-based geospatial problem-solving architecture.

Parallel Algorithm Design. Parallel algorithms for GeoComputation play a fundamentally significant role in the Grid-based geospatial problem-solving architecture to fully utilize the high-performance computing power of the Grid.

Due to GeoComputation's nature of high diversity, GeoComputational processes vary in data requirements, operational scope, and more importantly, workflow. It might be very hard for a generic parallel GeoComputation library or middleware to fully utilize the parallel computing resources across the Grid for all different kinds of GeoComputation. If a parallel GeoComputation library is to be developed, it should include special parallel algorithms for different GeoComputational processes.

Fortunately, CA models were born to be parallelized! The transition rules are applied to every cell in the cell space independently. From the perspective of parallel algorithm design, the whole cell space, which is usually stored in raster data format, is easy to decompose into sub cell spaces and assign onto multiple computing units across the Grid, e.g., processors. Then the transition rules can be applied on these sub cell spaces simultaneously.

However, Geo-CA models have their distinguishing characteristics. A cell space in a Geo-CA model seems to be regular in terms of cell size and cell locations, but cells' properties are usually heterogeneous over the space. For example, in a cell space representing urban areas, urbanized cells might be concentrated in some parts of the space, but sparse in the others. Also, there might be some areas exclusive for urban sprawl, e.g., oceans, national parks. Simply dividing the cell space into equal-area sub cell spaces will cause inequality of sub cell spaces' workloads which are the numbers of non-urbanized cells in non-exclusion areas. To solve this workload inequality, a spatially adaptive decomposition method has to be deployed. One approach is to decompose the cell space into unequal-area sub cell spaces according to the workloads associated with them (Fig. 2). Spatial indexing techniques, which are usually used in geospatial databases, could be used to produce workload-oriented spatially adaptive decompositions. A quad-tree-based domain decomposition approach has been implemented to support a Grid-enabled parallel spatial interpolation algorithm [8]. Furthermore, in a computational Grid environment, due to the high diversity of computing resources and the variability of network communication rates, reducing the granularity of decomposition, and assigning multiple sub cell spaces onto a computing unit according to its computing capacity and status (busy/idle) and the current status of the network, will be useful to deliver the appropriate amount of workload to the computing unit.

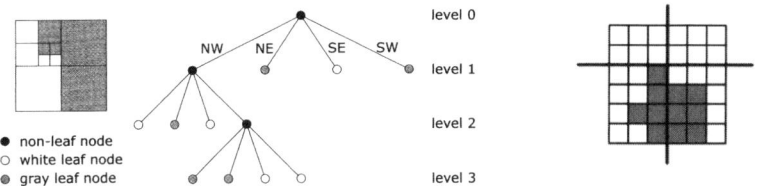

Fig. 2. The region quad-tree [9], and a workload-oriented spatially adaptive decomposition where the numbers of white cells are the workloads

Another issue involved in a parallel Geo-CA is the neighborhood. Since the transition rules applied in a specific cell require its neighborhood's data, a sub cell space assigned to a computing unit has to have overlapping belt(s) with its neighboring sub cell space(s). These overlapping belts, referred as "ghost cells", are actually copies of corresponding cells in the neighboring sub cell space(s). At the end of each transition process, the ghost cells' states have to be updated according to the new states of their origins which are usually owned by other computing units. This creates communication overheads among computing units. In a Geo-CA model, not all the cells change their states at each iteration. To minimize the communication overhead, only those "ghost cells" whose origins' states have been changed need to be updated. We call this approach "update on change".

To facilitate this "update on change", an object-oriented data model was designed to ensure effective and efficient communications among computing units (Fig. 3). The *CellSpace* class contains some basic information about the cell space, and a matrix of cells that are associated with states from a predefined set of states. The *count* function of the *CellSpace* class, which calculates the number of cells that have certain states, e.g., non-urbanized cells, is to provide an estimate of the cell space's workload. The *SubCellSpace* class, derived from the *CellSpace* class, has all the properties and functions of the *CellSpace* class and a few additional ones. The *nbrSpcIDs* is an array of the neighboring sub cell spaces' IDs. The *sendRange* stores the location ranges of cells that need to be sent to the neighboring sub cell spaces. The *cells2Send* stores the arrays of cells to be sent to the neighboring sub cell spaces at the end of each iteration. Similarly, the *cellsRcved* stores the arrays of cells received from the neighboring sub cell spaces that are actually the ghost cells to be updated. Note that both *cells2Send* and *cellsRcved* store arrays of "*globally indexed cells*". A globally indexed cell consists of a state value and a global index to ensure that it will be correctly recognized and located across sub cell spaces. Also, a global index is smaller than a row-column coordinate pair in terms of data volume and benefits efficient storage and transfer. A global index of a cell can be computed using the *lclCoord2glbIdx* function, and translated into a local coordinate pair using the *glbIdx2lclCoord* function.

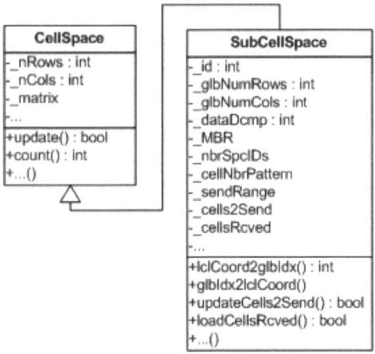

Fig. 3. An object-oriented data model for parallel Geo-CA models

Implementation on the Grid. This parallel Geo-CA model has been implemented using C++ language. In order to integrate the Geo-CA model into the Grid-based geospatial problem-solving architecture, several issues have to be concerned. First, this parallel Geo-CA model resides in the *Model* tier of the architecture, which links to the *Service* tier and the *Grid* tier. The program is stored in a computer on the Grid. Then an OGC-compatible GIService, serving as the interface between the web portal and the model, needs to be developed and published on the Grid. Through the web portal, end users are able to custom the Geo-CA model, e.g., setting parameters and transition rules, specifying input data, even integrating the Geo-CA model into a bigger model to simulate a complex geospatial phenomenon and solve a particular geospatial problem. The Resource Management and Allocation module in the *Grid* tier automatically locates the available computing resources, e.g., processors on the Grid, and data depositories, and forms a "virtual parallel computer" for the parallel Geo-CA program. The Geo-CA program decomposes the task into sub tasks and assigns them onto multiple computing units according to their computing capacities and the network status. The user is able to monitor the status of the program and stop it if something goes wrong. The results will be returned to the GIService and displayed to the user in the *Presentation* tier.

5   Conclusions

GeoComputation implies the revolutionary application of computational science in geography or an even broader domain. After a decade of research and development, GeoComputation faces some challenges that can not be tackled using traditional computing technologies. These challenges are mainly introduced by GeoComputation's nature of high diversity, the increasing complexity of geospatial models and the ever-growing amount of high-resolution geospatial data. New computing technologies, especially Grid Computing, offer an opportunity to promote GeoComputation in terms of usability, feasibility, applicability and availability. A Grid-based geospatial problem-solving architecture proposed in this paper provides a solution for building an easy-to-use, widely accessible and high-performance geospatial problem-solving environment integrating multiple complicated GeoComputational processes at an acceptable cost. A parallel geographic cellular automata model is given as an example to address some distinguishing issues when designing and implementing Grid-enabled parallel GeoComputation algorithms and models to effectively and efficiently utilize the computational Grid. A workload-oriented spatially adaptive data decomposition strategy is needed to deliver the appropriate amounts of workloads to computing units across the Grid. The "update on change" approach will reduce the communication overheads and object-oriented data models are useful for facilitating it. An OGC-compatible GIService needs to be developed to integrate the parallel GeoComputation model into the Grid-based geospatial problem-solving architecture, and the Grid Resource Management and Allocation module is critical for forming a "virtual parallel computer" for the model.

  Our future work includes developing more parallel GeoComputation algorithms and models for different geospatial applications, and implementing them into Grid-compatible GIServices to build a Grid-based web portal for an integrated geospatial problem-solving environment.

References

1. Couclelis, H., *Geocomputation in Context*, in *GeoComputation: A Primer*, P.A. Longley, et al., Editors. 1998, Wiley: New York. p. 17-30.
2. Foster, I., C. Kesselman, and S. Tuecke, *The Anatomy of the Grid: Enabling Scalable Virtual Organizations*. International Journal of Supercomputer Applications, 2001. **15**(3).
3. Openshaw, S., *GeoComputation*, in *GeoComputation*, S. Openshaw and R.J. Abrahart, Editors. 2000, Taylor & Francis: New York. p. 1-31.
4. Baker, M., R. Buyya, and D. Laforenza, *Grids and Grid Technologies for Wide-area Distributed Computing*. Software - Practice and Experience, 2002. **32**(15): p. 1437-1466.
5. Foster, I., *What is the Grid? A Three Point Checklist*. Grid Today, 2002. **1**(6).
6. Foster, I. and C. Kesselman, *Computational Grids*, in *The Grid: Blueprint for a Future Computing Infrastructure*. 1999, Morgan Kaufmann Publishers: San Francisco, CA.
7. Clarke, K.C., *Geocomputation's Future at the Extremes: High Performance Computing and Nanoclients*. Parallel Computing, 2003. **29**(10): p. 1281-1295.
8. Wang, S. and M.P. Armstrong, *A Quadtree Approach to Domain Decomposition for Spatial Interpolation in Grid Computing Environments*. Parallel Computing, 2003. **29**(10): p. 1481-1504.
9. Worboys, M. and M. Duckham, *GIS: A Computing Perspective*. 2nd ed. 2004, New York: CRC Press.

Semantic Spatial Web Services with Case-Based Reasoning

Taha Osman[1], Dhavalkumar Thakker[1], Yanwu Yang[2,3], and Christophe Claramunt[2]

[1] School of Computing and Informatics, Nottingham Trent University, Nottingham, UK
{Taha.Osman, Dhavalkumar.Thakker}@ntu.ac.uk
[2] Naval Academy Research Institute, BP 600, 29240 Brest, France
{yang, claramunt}@ecole-navale.fr
[3] Institute of Software, Chinese Academy of Sciences, Beijing 100080, China
yangyanwu@ercist.iscas.ac.cn

Abstract. With the rapid proliferation of spatial information on the web, the development of spatial web services represents a challenging issue. The research presented in this paper introduces a novel approach for the integration of spatial semantics within a case-based reasoning solution for the delivery of spatial information services on the Web. The framework integrates Web services in the decision making process and is adaptable to service requester constraints. The framework is based on OWL semantic descriptions for implementing both the components of the CBR engine and the matchmaking profile of the Web services. The framework and approach are illustrated with web-based travel planning, e.g. flight schedule arrangement according to user's requests, constraints, and preferences.

1 Introduction

Automatic Web service discovery and matchmaking is an important aspect of dynamic services composition. The accuracy of a matchmaking process enhances the possibility of successful composition, eventually satisfying the user and application requirements. The current standard for Web service discovery, the Universal Description, Discovery and Integration (UDDI) registry, is syntactical and has no scope for automatic discovery of Web services. Hence, current approaches attempting to automate service discovery and matchmaking processes apply different sorts of semantics to service descriptions. These semantics are interpretable by the service agents and include WSDL-based functional parameters such as Web services input-outputs [1][2], and non-functional parameters such as domain-specific constraints and user preferences [3].

The accuracy of automatic matchmaking web services can be improved by taking into account the adequacy of past matchmaking experiences for a requested task. This gives valuable information about services behaviors that are difficult to presume prior to service execution. However, there is a still a need for a methodology that uses domain-specific knowledge representation of the required tasks to capture the Web services execution experiences and use them in matchmaking processes. Case Based Reasoning (CBR) provides such methodology as its fundamental premise is that

J.D. Carswell and T. Tezuka (Eds.): W2GIS 2006, LNCS 4295, pp. 247–258, 2006.
© Springer-Verlag Berlin Heidelberg 2006

experience formed in solving a problem situation can be applied to other similar problems.

In a related work we introduced a research framework that integrates CBR methodology and semantic descriptions to enhance web services discovery [4]. We adopted CBR as the engine for Web services discovery mechanism as CBR's fundamental premise is that situations recur with regularity, i.e. experience involved in solving a problem can be applied or can be used as guide to solve other similar problems [5]. A reasoner based on CBR matches previous experiences to derive a solution for new problems [6]. The CBR engine is employed to generate web services taking into account past service execution experiences and services. Request cases are represented using a web ontology language OWL, itself based on a domain ontology.

The motivation of the research presented in this paper is that a significant proportion of Web resources can be associated to some degree to geo-referenced entities. Statistics collected by search engines and systems on the Web show that spatial information is pervasive on the Web, and that many queries explicitly or implicitly contain spatial criteria. Our objective is to introduce a degree of spatial semantics in CBR case retrieval. We introduce and develop a model which is illustrated with web-based travel planning, e.g. flight schedule arrangement, according to user's requests, constraints, and preferences.

The rest of the paper is organized as follows. Section 2 introduces the principles of Web services matchmaking. Section 3 presents the spatial and hierarchical-based measures integrated in the case retrieval process. Preliminary implementation of the framework is described in section 4. Finally, section 5 concludes the paper.

2   Web Services Using Case Based Reasoning

2.1   The Framework Architecture

The main roles in Semantic CBR matchmaking are the ones of the case administrator who is responsible for case library maintenance by entering or deleting cases from the library, and the one of the case requestor who searches for the case library to find a solution for a given problem. Figure 1 illustrates a schematic diagram of the framework. A detailed explanation of the framework dynamics is given in a related paper [4].

2.2   Ontology Support for Case Representation and Storage

One of the most common uses of ontologies is the reconciliation between syntactically different terms that are semantically equivalent. Applied to CBR case descriptions for Web services, ontologies can be used to provide a generic, reasoner-independent description of their functional and non-functional parameters. Ontologies can also be used to index and structure the different cases with key domain features that increase the efficiency of the matchmaking process. For instance, one can add a feature to the travel domain ontology to indicate whether a trip is domestic or international. Web services parameters are also indexed using ontologies to improve the accuracy of case matchmaking. In the framework, ontologies support the description   of the CBR reasoning engine, which not only streamlines the intercommunication between the Web service, user request, and the case library, but

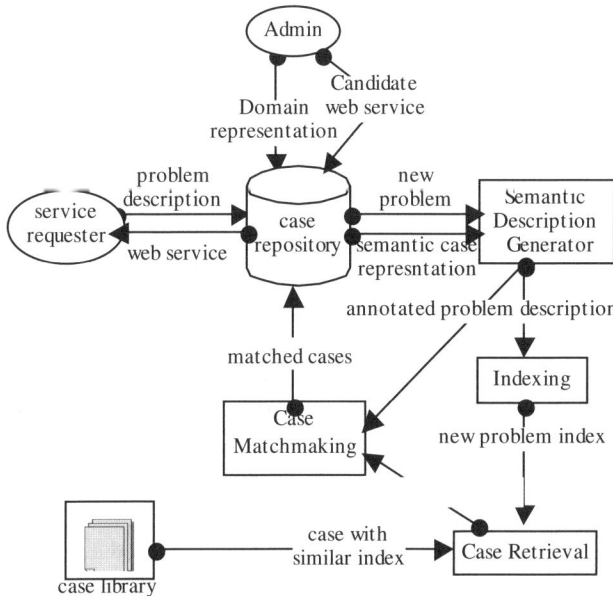

Fig. 1. Architecture of the CBR matchmaking framework

also promotes exploring the collaboration at the reasoning level between different composition frameworks.

2.2.1 Case Vocabulary

In CBR modeling, the first step is to define all the elements contained in a case and the associated vocabulary that represents the knowledge associated with the context of a specific domain (e.g. the travel domain in our case study). This vocabulary includes functional and non-functional parameters:

1. Functional parameters are the service inputs (e.g. travel details), and the service outputs (e.g. travel itinerary). Inputs correspond to the user request (e.g. date or city of departure) whereas outputs correspond to the response given to the user (e.g. flight schedule).

2. Non-functional parameters are constraints imposed by the user (e.g. exclusion of particular travel medium) or preferences over certain specific parameters (e.g. price range, quality of Service expected). In addition, execution experiences stored in the case library should also include the solution (i.e. Web services effectively used) and a notion to specify if the solution is acceptable for the end-user. Features that characterise the domain are extremely useful for top-level indexing and can also be included as non-functional parameters.

2.2.2 Case Representation Using Frame Structures

Our approach retains frame structures [7] for case representation. In frame structures, a frame is the highest representation element consisting of slots and fillers. Slots have dimensions that represent the lower level elements of the frame, while fillers are the value range the slot dimensions can draw from. In our implementation, slot dimensions represent the case vocabulary in a modular fashion while fillers describe the possible value ranges for the slot dimensions.

Table 1. Case representation

Slot	Dimension	Filler
Travel Request	Name of Traveler	Any text
	Date of Arrival	Any valid date
	City of Departure	Any valid city
Travel Response	Solution	Service WSDL file
	Price Range	Any positive Double
	Currency	Any valid currency
Constraints	On Domain	Any Valid Travel Domain
	On Price range	Any positive Double
	On QoS parameter	Any possible QoS parameter(s)
Features	Travel Regions	Domestic/International

The frame representations are highly structured and modular, and have a direct mapping to the semantic OWL description language as the semantic net represent-tations largely borrowed from frame structures [8], which makes natural transition to the Semantic Web descriptions possible. Table 1 shows such an example of frame structure of the travel domain case vocabulary.

The slot Travel Request corresponds to the inputs, i.e. all the travel details as for any travel agent. The Travel Response slot corresponds to the Output, i.e. the answer given to the user at the end of the process. The elements of the answer are the price and the corresponding currency, the access point to the WSDL file of the corresponding Web Services and the Services Used (e.g. companies involved in the trip).

2.2.3 Semantic Encoding of the Frame Structure

The frame structures can be mapped to ontologies as described in Figure 2. According to this mapping, *frame* and *slot* are represented as classes. The relationship between *frame* and *slot* is expressed in terms of properties of *frame*, as the range for these properties are the *slot* classes. *Dimensions* are the properties of the *slots*. Possible range for these properties is the values the respective *filler* can derive from.

We use Web Ontology Language (OWL), a Semantic Web standard for constru-cting these ontologies. The layered approach adopted by the semantic web, allows reasoning and inference based on ontologies, which is an ubiquitous feature of the Semantic Web. After applying the mapping, the ontology for the travel domain the case representation is created, where for instance the *CaseRepresentation* class has: *hasTravelResponse*, *hasConstraintsOnGoal*, and *hasFeature* object properties. Range for these properties are *TravelResponse*, *Constraints*, and *Feature* classes, respectively. In order to genralise the semantic descriptions, external ontologies are used where appropriate. For instance, the property *cityOfArrival* is an object property referring to a publically available ontology [9], where other useful city information.

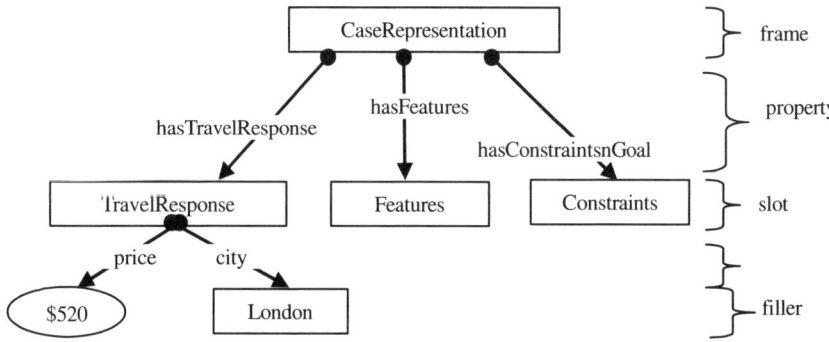

Fig. 2. Mapping between frame structure and semantic case representation for a travel domain

All the Web service execution experiences, i.e. solutions deemed valid for a particular request, are stored in the Case Library to be reused by the reasoner. The Case Library itself is also an ontology, it contains some instances of the class CaseRepresentation (e.g. a travel experience or a travel case).

3 Development of the CBR Framework

3.1 Case Indexing and Retrieval

In order to facilitate the search procedure, the indexation of the cases are based on vocabularies, it is a variation of the "flat memory indexing" technique [7]. A novel problem is tentatively recognized based on the identical vocabularies to decide which partition the problem falls into.

Cases are stored based on vocabulary element Features as presented in Table 1, which corresponds to *hasFeatures* property (see Figure 2) from the *CaseRepresentation* ontology class. Regarding the travel agent case study, the possible values for this property are either Domestic or International (predefined instances from the TravelRegion class), hence indexing will partition the case library into two parts. The efficiency of the retrieval process largely depends on the precision of the indexing. Whenever a new Web service needs to be fetched, the problem description involving the functional parameters and non-functional parameters are encoded using the case representation frame structure, i.e. as an instance of *CaseRepresentation* ontology as illustrated in Table 2.

3.2 Matchmaking and Ranking

Case retrieval fetches Web services that are a potential solution to the problem. The matchmaking process narrows down the retrieved cases to present acceptable solution(s). From current matchmaking methods available in CBR, nearest-neighbour matching combined with a ranking numeric evaluation function [10] are used as references for our framework. The method operates as follows:

1. Compare the similarity for each property, between the new problem and the cases retrieved. The method used for comparison depends on the type of the property;

2. Quantify the weight of the similarity. A ranking is assigned to each property in accordance with its importance as exemplified in Table 2.

Table 2. Quantifying the Travel Domain case dimensions

Slot	Dimension	Importance [0-1]
Travel Request	City Departure	1.0
	City Arrival	1.0
Constraints on Goal	On Instance	0.2
	On Domain	0.8

For each case retrieved, the similarity degree is computed and the case with the highest score corresponds to the best-match. Similarity takes values between 0 and 1, which is attributed to each property for each retrieved case. The similarity comparison method depends on the type of the dimension: *data* or *object*. A special kind of entity, *spatial entity*, is considered, e.g., to measure the closeness between two spatial entities involved in case match-making processes.

A case's relative utility and feedbacks refer to, to which degree it is either useful for a given problem or not. Case's utility is usually evaluated with certain similarity measures. The definitions of accurate and flexible similarity measures are a key issue in CBR applications [11]. Current approaches mostly employ either a statistical analysis of the case base, or user's relevance feedbacks to evaluate the accuracy of case retrieval results. However, these approaches are restricted to simple classification tasks, many application domains require more accurate knowledge-intensive similarity measures [12]. The knowledge-intensive similarity measures use domain knowledge to approximate case's utility. Applied to our application context, a travel case can be mapped to a geo-referenced event in the underlying physical environment. This implies to combine spatial and semantic criteria in case retrieval and similarity measures.

3.2.1 Data Property Comparison

In order to compare data type properties, like the price range or the value of QoS (e.g. execution time), a region-based measurement method is employed [7]. The closer the value in a retrieved case is to the value in the request, the higher the similarity coefficient is.

For each data type property, the formulae used is: $|V_r - V_c| \leq X.[V_r|$, where V_r is the value of the property in the request r, V_c in the retrieved case c , and X the factor of tolerance. For example, a factor of tolerance of 0.9 means that the value of the retrieved case should be in $\pm 10\%$ region in relation to the value of the request. The optimum tolerance value is determined by the administrator and can be calculated heuristically.

3.2.2 Object Property Comparison --- Hierarchical Similarity Measure

Regrding the *dimensions* annotated as object properties, the possible *filler* values are an instance of *slot* class. For semantically matching object property value of the new problem and the retrieved cases, the algorithm compares the instances. If the instances match, then the degree of match is 1. Otherwise, the algorithm traverses back to the super (upper) class that the instance is derived from and the comparison is performed at that level.

The hierarchical comparison is similar to traversing a tree structure [13], where the tree represents the class hierarchy for the ontology element. The procedure of traversing back to the upper class and matching instances is repeated until there are no super classes in the class hierarchy, i.e. the root node for the tree is reached, giving degree of match equal to 0.

We adopted a hierarchical similarity measure that combines the commons and differences to determine the closeness between two nodes, and taking into account contextual knowledge in a given hierarchy [14]. An hierarchical domain ontology consists of a set of semantic classes N and links L. Classes are labeled with distinct labels. Links connect classes with different relationships e.g. is-a and part-whole. Let H be a hierarchical domain, Root (H) the root. The depth of a class is the number of links between Root (H) and the class. The least common ancestor of two classes is the deepest subsumer of them. The relationships between two semantic classes can be represented either by the number of links connecting them in the hierarchical structure, or by a function of the number of their common and distinctive super classes. The links and classes are also assigned weights denoting different importance, based on depth and density of semantic classes in class hierarchy.

Let $sup(C_1)$ be the set of super classes of C_1 in the hierarchical domain ontology,
   $deep(C_1)$ the depth of C_1,
   $sib(C_1)$ the number of siblings of C_1 with the most specific, common ancestor,
   $sup(C_1/C_2)$ the set of super classes of C_1 but not of C_2,
   $dis(C_1, C_2)$ the number of links between C_1 and C_2,
   $LCA(C_1, C_2)$ the least common ancestor of C_1 and C_2.
The similarity between two semantic classes C_1, C_2 in a given hierarchical domain ontology is given as follows [14]:

$$sim(C_1, C_2) = \frac{\lambda \mid sup(C_1) \cap sup(C_2) \mid}{\mid sup(C_1) \cup sup(C_2) \mid + \alpha \mid sup(C_1/C_2) \mid - (1-\alpha) \mid sup(C_2/C_1) \mid} \tag{1}$$

Where α is a parameter bounded by the unit interval [0, 1], λ the depth parameter.

The weight α is determined as a function of the distance between semantics C_1, C_2 and the least common ancestor of both classes, and the number of sibling C1, C2. It is given as

$$\alpha(C_1, C_2) = \frac{dis(C_1, LCA(C_1, C_2)) \times sib(C_1)}{dis(C_1, LCA(C_1, C_2)) \times sib(C_1) + dis(C_2, LCA(C_1, C_2)) \times sib(C_2)} \tag{2}$$

The depth parameter λ is given as,

$$\lambda = \frac{2deep(LCA(C_1, C_2))}{deep(C_1) + deep(C_2)} \tag{3}$$

The similarity function yields values bounded by the unit interval [0,1]. The maximum value 1 occurs *iff* the two semantic classes under comparison are equivalent, that is, $C_1=C_2$. The similarity function reflects an asymmetric relationship between two semantic classes.

3.2.3 Proximity Measures

Proximity between spatial entities is a fundamental relationship when considering the case of a user physically acting and moving in a given physical environment [15]. We integrate a spatial proximity measure that considers a hierarchical-based notion of spatial proximity to evaluate the closeness between two spatial entities (e.g. departure city and arrival city) involved in a request case and a candidate case respectively. This spatial proximity measure takes into account the fact that *the semantics difference between two entities increases with the decrease of the distance between them.* The geometrical spatial proximity reflects user-centric principles observed in qualitative studies [16] [17], that *the distance from a spatial entity α to a distant entity β should be magnified when the number of regions near α increases, and vice versa.* The hierarchical spatial proximity is computed with similarity measure (as given in section 5.2.3) in a hierarchical structure of world cities with *part-of* relations (Figure 3).

Geometrical spatial proximity

The *contextual distance* normalizes the conventional Euclidean distance from a spatial entity α to another β by a dividing factor that gives a form of contextual value to that measure (4). The dividing factor is given by the average of all distances between α and other spatial entities. In this context, it is given as a homogeneous form of "contextual distance" defined in   [18], so is the contextual proximity. The *contextual distance* between two entities x_i and x_j of $X=\{x_1, x_2, ..., x_p\}$ is given by

$$GS\Pr ox(x_i,x_j)=\frac{\overline{d(x,X)}^2}{\overline{d(x,X)}^2+d(x_i,x_j)^2} \tag{4}$$

where $d(x_i, y_j)$ stands for the Euclidean distance between x_i and y_j; $\overline{d(x,X)}$ the average distance between x_i and the other entities of X. The higher $GS\Pr ox(x_i,x_j)$ the closer x_i to y_j, the lower $GS\Pr ox(x_i,x_j)$ the distant x_i to x_j.

Hierarchical spatial proximity

Spatial entities can be organized using a hierarchical structure according to classes that denote natural or man-made boundaries. A spatial hierarchy is organized with part-of relationships between classes. Proximities in the hierarchical structure complement the proximities in the spatial dimension. The proximity extracted from such a spatial hierarchy is termed "hierarchical spatial proximity". It can be computed with the hierarchical similarity measure introduced in section 3.2.2, that is, $HSProx(x_i,x_j)=sim(x_i,x_j)\cdot$

Overall spatial proximity measure

Given the hierarchical spatial proximity computed from a hierarchical representation of spatial entities with the part-of relation (Figure 3) with equation (1), and geometrical spatial proximity with equation (4) between a spatial entity (in a request case) and the corresponding entity (in a candidate case), the overall spatial proximity is given as,

$$SProx(x_i, x_j) = \eta GS\Pr ox(x_i, x_j) + (1 - \eta)HSProx(x_i, x_j) \tag{5}$$

Where The constant η is valued as a real number between 0 and 1 to moderate the respective influence of the geometrical and hierarchical spatial proximities. It is valued as 0.2 as two travel cases that start (or stop) in a same city/country/continent are likely to be more similar than the ones in different counties.

3.2.4 Overall Similarity Value

The overall similarity is evaluated by computing the aggregate degree of match (ADoM) [10] for each retrieved case according to the following equation:

$$ADoMSim(N, R) =$$

$$\frac{\sum_{i=1}^{n-2} W_i \times sim(f_i^N, f_i^R) + W_{n-1} \times S\Pr ox(f_d^N, f_d^R) + W_n \times S\Pr ox(f_a^N, f_a^R)}{\sum_{i=1}^{n} W_i} \tag{6}$$

Where, n is the number of ranked dimensions, W_i is the importance of dimension i, sim is the similarity function for general non-spatial primitives, and f_i^N and f_i^R are the values for feature f_i in the new problem N and the retrieved case R respectively. $S\Pr ox(f_d^N, f_d^R)$ and $S\Pr ox(f_a^N, f_a^R)$ denote the overall spatial proximity between departure/arrival cities involved in case matching, respectively. The spatial proximity between departure cities, and the one between arrival cities act as two special dimensions in the aggregate function. The evaluation function sums the degree of match for all the dimensions as computed in step A, and takes aggregate of this sum by considering the importance of the dimensions.

4 Implementation: A Travel Case Retrieval Example

The implementation takes into account the ontologies describing the components of the Case-Based Reasoner (Case representation), and the domain ontologies that describe the profile of the Web services in the Case library with a semantic representation (Case Storage). The implementation is based on Pellet [19] - a Java based OWL reasoner, together with Jena [20] that supports user-defined simple types. Pellet was used to load and verify (type and cardinality) ontology class instances of user requests and candidate cases. While seeding the case library with a new case or making a new trip request, the system assists the client in creating the required ontology instances. The value entered for a particular property is validated in relation to the range and cardinality drawn from the ontologies. The solutions resulting from the matchmaking process are presented to the client and stored into the case library.

At this initial stage of our development, our experiments focus on the validation of the logic of our CBR retrieval and matchmaking framework, rather than testing a fully working prototype. Hence, we tested our framework with simple in-house developed Web services and compatible wrappers for external publicly available services.

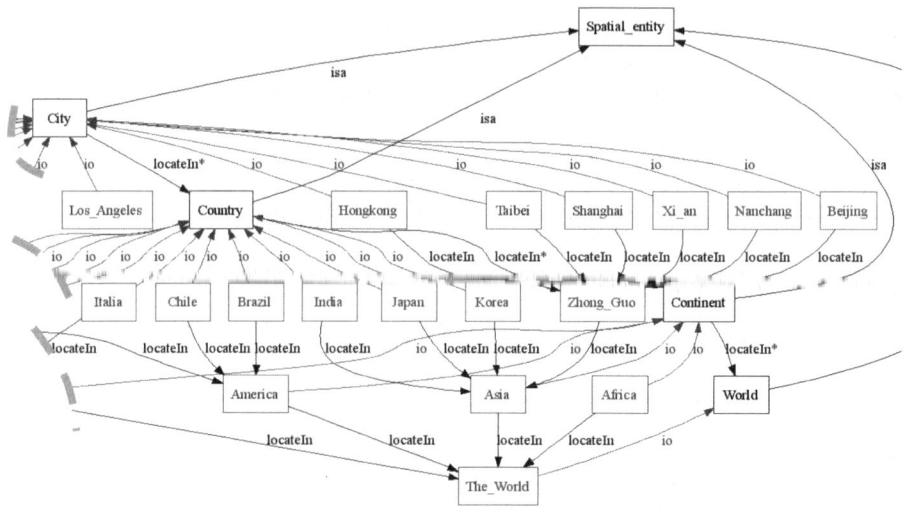

Fig. 3. Spatial hierarchy (partial view)

Let's consider a travel request from a given user: "I want to leave from Brest to Xi'an, July 21, 2006, and arrive in the same day. I prefer to take plane, and to pay in Euro, and less than 750 euros is acceptable... " In order to illustrate such as a case retrieval and matchmaking, we use five candidate cases in our case library: case#1: Hongkong → Los Angeles; case#2: London → Paris; case#3: Paris → Beijing; case#4: Lyon→Hongkong; case#5: Paris→Nanchang. Each candidate case is described with relevant criteria as defined in section 2.2.

The case matchmaking and similarity assessment begin with a computation of the proximity values in each individual dimension. The spatial proximity is considered with respect to two possible dimensions: city departure and city arrival. The geometric spatial proximity (e.g. between cities in our context) is computed with positional data, and taking into account the overall spatial distribution of the entities available in the case library. The hierarchical spatial proximity between the spatial entities involved in our example is computed from a spatial hierarchical representation of world cities partially illustrated in Figure 3.

The case matching process starts with the computation of the spatial proximity with respect to the departure and arrival city dimensions, and then with the hierarchical ontology. These computational outputs reflect the closeness between user's request case and each candidate case. Proximity values are aggregated using equation 6 with weights assigned according to user's implicit or explicit requirements. In this case retrieval example, besides departure city and arrival city (with weights assigned as 1), the weights for different dimensions are valued as follows: domain (weight =1), price (weight=0.8) and QoS (weight=0.1). The final similarity results between the request and the three candidate cases turns out that case#3 is the most similar with respect to user's request. It is worth mentioning that, given the fact that the other criteria are similar, Beijing is more distant from Xi'an than Nanchang is, in the Euclidean space. However it's different according to the contextual and hierarchical proximity. The reason behind is that Nanchang is located not far away from several cities available, such as Taibei, Hongkong and Shanghai. This illustrative solution is presented to the user at the interface level. If the user is not completely satisfied with it, then a series of tweaking actions will be activated to adapt the travel plan to user's specific needs.

5 Conclusion

Semantic description of Web service profiles paves the way for automating the discovery and matchmaking of services since it allows intelligent agents to reason on service parameters and capabilities. However, the accuracy of such automatic search mechanisms largely relies on how soundly formal methods working on such semantic descriptions consume them.

This paper introduces a Semantic Case based Reasoner, which captures Web service execution experiences as cases and uses them for deriving a solution for new problems. The implemented framework uses ontologies, as semantics are used for describing the problem parameters and for implementing components of the CBR system: representation, indexing, storage, matching and retrieval. The CBR approach is extended by an ontological component and spatial semantics-based measures that facilitate reasoning capabilities. The hierarchical- and semantics-based proximity measures take into account contextual knowledge extracted from the overall distribution and naturally (instead of strictly) hierarchical, domain-dependent representation of spatial entities. The preliminary experimental results of this framework are presented with a travel case retrieval example.

Future work will involve exploring case adaptation and personalization services, which are applicable when the available cases cannot fulfil the problem requirements. Adaptation is similar to Web service composition, as the composition is applied when available services are not sufficient in meeting the requirement for the problem. On the other hand, it is an interactive process with the user, through which user's interests and preferences can be inferred for the generation of web services tailored to user's specific needs.

References

[1] Martin, D. *et al.*, 2004, Bringing Semantics to Web Services: The OWL-S Approach, in Proceedings of the First International Workshop on Semantic Web Services and Web Process Composition (SWSWPC 2004), San Diego, USA.

[2] Akkiraju, R., Farrell, J., Miller, J., Nagarajan, M., Schmidt, M., Sheth, A. and Verma, K., 2005, Web Service Semantics - WSDL-S, UGA-IBM Technical Note, version 1.0.

[3] Aggarwal, R., Verma, K., Miller, J.A. and Milnor, W., 2004, Constraint Driven Web Service Composition in METEOR-S, in Proceedings of the IEEE International Conference on Services Computing (SCC 2004), Shanghai, China, pp. 23-30.

[4] Osman, T., Thakker, D., Al-Dabass, D., Lazer, D. and Deleplanque, G., 2006, Semantic-Driven Matchmaking of Web services using Case-Based Reasoning, in Proceedings of the IEEE International Conference on Web Services (ICWS 2006), September 18-22, Chicago, USA, to appear.

[5] Aamodt, A. and Plaza, E., 1994, Case-based reasoning: foundational issues, methodological variations, and system approaches, AI Communications, 7(1), 39-59.

[6] Watson, I., 1997, Applying Case-Based Reasoning: Techniques for Enterprise Systems, Morgan Kaufmann Publishers, San Francisco, 285 p.

[7] Kolodner., J. and Simpson, R., 1989, The MEDIATOR: Analysis of an Early Case based Problem-solver, Cognitive Science, 13(4), 507-549.

[8] Rich, E., and Knight, K., Artificial Intelligence. McGraw-Hill, 1992.

[9] Portal Ontology. AKT Technologies. 10 Feb 2003. Available at: http://www.aktors.org/ontology/portal#

[10] ReMind Developer's Reference Manual, 1992, Cognitive Systems, Boston.

[11] Armin, S., 2002, Defining Similarity Measures: Top-Down vs. Bottom-Up, in Proceedings of 6th ECCBR conference, Springer LNCS 2416, pp. 406-420.

[12] Armin, S. and Gabel, T., 2006, Optimizing Similarity Assessment in Case-Based Reasoning, in Proceedings of the 21th National Conference on Artificial Intelligence (AAAI-06), AAAI Press.

[13] Zhang, R., Budak I. and Aleman-Meza, B., 2003, Automatic Composition of Semantic Web Services, in Proceedings of the International Conference on Web Services, ICWS '03, Las Vegas, Nevada, USA., pp.38-41

[14] Yang. Y. 2006, Towards Spatial Web Personalization, Unpublished PhD report, Naval Academy Research Institute, France, 158p.

[15] Tobler, W.R., 1970, A Computer Model Simulating Urban Growth in the Detroit Region. Economic Geography, 46, 234-240.

[16] Sadalla, E.K., Burroughs, W.J. and Staplin, L.J., 1980, Reference Points in Spatial Cognition. Journal of Experimental Psychology: Human Learning and Memory, 5, 516-528.

[17] Tversky, B., 1993, Cognitive Maps, Cognitive Collages, and Spatial Mental Models. In Frank, A.U. and Campari, I. (eds.), Spatial Information Theory, Proceedings of COSIT'93, Springer LNCS 716. Berlin, Germany, pp. 14-24.

[18] Yang, Y. and Claramunt, C., 2004, A Flexible Competitive Neural Network for Eliciting User's Preferences in Web Urban Spaces, in Fisher P (eds.), Developments in Spatial Data Handling, Springer-Verlag, pp. 41-57.

[19] Parsia, B., and Sirin, E., 2004, Pellet: An OWL DL Reasoner, in Proceedings of the Third International Semantic Web Conference (ISWC2004), Hiroshima, Japan.

[20] Jena, 2006, A semantic Web Framework for Java, HP Labs Semantic Web Programme, http://jena.sourceforge.net/

TMOM: A Moving Object Main Memory-Based DBMS for Telematics Services

Joung-Joon Kim, Dong-Suk Hong, Hong-Koo Kang, and Ki-Joon Han

School of Computer Science & Engineering, Konkuk University,
1, Hwayang-Dong, Gwangjin-Gu, Seoul 143-701, Korea
{jjkim9, dshong, hkkang, kjhan}@db.konkuk.ac.kr

Abstract. Recently with the growth of the Internet and the activation of wireless communication, telematics services are emerging as promising next-generation business in the IT area. In order to provide telematics services, technologies in various areas are required but particularly DBMS technology for efficient data processing and management is an essential key technology for all types of telematics services. This paper designed and implemented TMOM (Telematics Moving Object Main Memory DBMS) to meet efficiently the requirements that telematics services should be provided in real-time. TMOM follows the spatial data model of OpenGIS "Simple Features Specification for SQL", and provides spatial, spatio-temporal and temporal types and corresponding operators for processing the moving object data. In addition, it supports optimized spatial, spatio-temporal and trajectory indexes in the main memory for fast search of large moving object data and provides a recovery function that can minimize disk input-output to maximize system performance. Also, for high transmission efficiency in data import/export between telematics applications and a back-end DBMS, it supports a compression function and a data caching function suitable for the characteristics of spatial data.

Keywords: Telematics, Moving Object, MMDBMS, OpenGIS, Index, Data Compression, Recovery, Data Caching.

1 Introduction

Telematics services, which can be applied to real-time traffic information service, route search service, rescue service (e.g. 911), etc., are next-generation information services for car drivers and passengers. To provide telematics services, the telematics system must deal with a large amount and various types of location data and sensor data. Car location data to be managed for telematics services should be moving object data, in which the location or shape of objects continues to change over time, rather than spatial data expressed in simple coordinates. Also, because most telematics services should be provided in real time, they need to support quick response to data requests using a Main Memory-based DBMS (MMDBMS) [6,10].

MMDBMS has been being researched and developed in the U.S. and Europe, and commercial MMDBMS vendors are developing and selling MMDBMS that can be used on PC or mobile devices. Representative MMDBMS's include TimesTen of

J.D. Carswell and T. Tezuka (Eds.): W2GIS 2006, LNCS 4295, pp. 259–268, 2006.

Oracle, eXtremeDB of McObject, and Dali of AT&T Bell Laboratory. However, these MMDBMS's support only a small number of spatial data types and spatial operators, and their completeness is low. When the research directions of related research institutes and universities are examined, most researches are focused on high-performance query processing technology for processing complicated queries, indexing technology optimized to main memory, recovery technology in preparation for system failure, duplication technology for non-stop function, etc. [3,4,8], but not many researches are being made on data types, operators and indexes for processing moving object data. Therefore, it is urgently necessary to research and develop MMDBMS that has these functions and, at the same time, can process and manage large moving object data.

TMOM(Telematics Moving Object Main Memory DBMS) proposed in this paper provides spatial, spatio-temporal and temporal data types and corresponding operators following the spatial data model of OpenGIS "Simple Features Specification for SQL" [7] to process large moving object data, and supports spatial, spatio-temporal and trajectory indexes optimized to main memory for the fast search of large moving object data. In addition, it provides a recovery function with minimized disk input-output to maximize system performance. Moreover, to enhance transmission efficiency in data import/export between telematics applications and a back-end DBMS, it supports a compression function and a data caching function suitable for the characteristics of spatial data. Therefore, TMOM can be a very useful moving object MMDBMS running on a data server to provide efficient telematics services.

The structure of this paper is as follows. Following Chapter 1 Introduction, Chapter 2 reviews related works including the background of this paper and the requirements of the moving object MMDBMS. Chapter 3 explains the structure and key technologies of TMOM, and Chapter 4 describes performance evaluation. Lastly, Chapter 5 mentions the conclusion and possible future work.

2 Related Works

This chapter reviews related works including the background of this paper and the requirements of the moving object MMDBMS.

2.1 Background

Standard spatial data types and operators are presented in OpenGIS "Simple Features Specification for SQL" [7]. Table 1 shows spatial data types and spatial operators of OpenGIS.

Table 1. Spatial Data Types and Spatial Operators

Geometry Types	Spatial Relation Operators	Spatial Analysis Operators
• Point • LineString • Polygon • MultiPoint • MultiLineString • MultiPolygon • GeometryCollection	• Equals • Disjoint • Touches • Within • Overlaps • Crosses • Intersects • Contains • Relate	• Distance • Intersection • Difference • Union • SymDifference • Buffer • ConvexHull

OpenGIS defines various spatial data types and operators including 7 spatial data types (Geometry types) and 16 spatial operators (Spatial Relation Operators and Spatial Analysis Operators).

Researches related to the index of moving objects are largely divided into three categories [5]. First, researches related to indexes for searching present and future location include R-tree, Quad-tree and TPR-tree. Second, researches for searching past location include 3DR-tree, HR-tree and MV3R-tree. Third, researches on past trajectory indexes for the movement of moving objects include STR-tree, TB-tree and CR-tree (Combined R-tree). Indexes for high cache performance to optimize the main memory include CSB+-Tree, pR-Tree and CR-Tree (Cache-conscious R-tree). CSB+-Tree, which is a variation of B+-Tree, removes pointers to child nodes except the pointer to the first child node of a node and stores the child nodes consecutively in the main memory to reduce cache failure in B+-Tree. pR-Tree, which is a variation of R-Tree, removes the coordinates of child MBR (Minimum Bounding Rectangle) overlapping with the coordinates of parent MBR to reduce cache failure in R-Tree. Also, CR-Tree is a variation of R-Tree that compresses MBR, which occupies the largest part of the index in R-Tree, and uses the compressed MBR as keys to reduce cache failure in R-Tree.

Representative checkpoint techniques for MMDBMS are transaction consistent checkpoint, action consistent checkpoint, and fuzzy checkpoint technique [1]. The transaction consistent checkpoint technique carries out checkpoint at the point when all going transactions have been completed, and the action consistent checkpoint technique carries out checkpoint when actions have been completed, assuming that a transaction is the continuity of actions on DBMS. In case of the fuzzy checkpoint technique, if a checkpoint takes place, the technique creates a process to carry out checkpoint and carries out checkpoint asynchronously to other transactions. Because this technique carries out checkpoint separately from other transactions, it is superior in performance to other checkpoint techniques.

2.2 Requirements of Moving Object MMDBMS

Technology for the moving object MMDBMS is a key technology for all telematics services. To provide telematics services, the moving object MMDBMS needs efficient data modeling for continuously changing data to reduce the cost of data update, and various operators to process frequently used query forms efficiently [3].

In addition, it needs indexing technology optimized to the main memory for the fast search of large moving object data. The index has a structure that can improve search performance in telematics services in consideration of the characteristics of moving object location data [4]. What is more, in MMDBMS, most tasks are performed on the main memory but logging for recovery and checkpointing require disk input-output. Therefore, high-performance transaction processing technology and efficient recovery technology are required to maximize system performance through minimizing disk input-output in logging and checkpointing, which occupies most of transaction execution time [6].

Lastly, data compression technology is necessary for compressing and decompressing data for efficient network transmission in data import/export among various systems including telematics applications and middleware. What is more, in order to reduce transmission delay and raise data transmission rate in responses to users' queries, data caching technology is needed to send important data first and less important data later [10].

3 TMOM (Telematics Moving Object Main Memory DBMS)

This chapter explains the structure of TMOM and main technologies used in TMOM.

3.1 Structure of TMOM

Figure 1 shows the overall structure of TMOM.

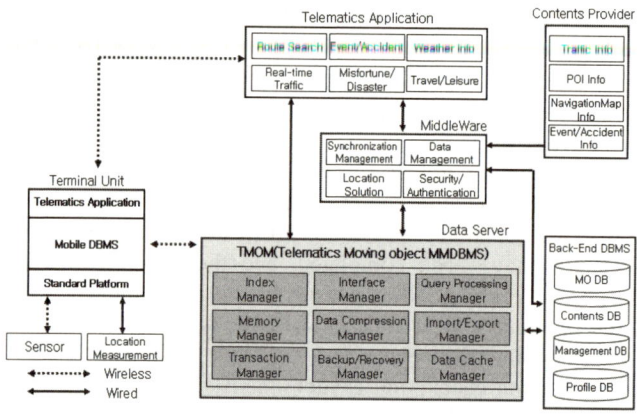

Fig. 1. Structure of TMOM

The interface manager provides a standard interface for various systems including telematics applications and middleware to access TMOM, and the query processing manager analyzes and processes queries using a query language compatible with SQL 99 such as DDL, DML and DCL, and executes search, insert, delete and update operations on various types of data using spatial, spatio-temporal and temporal data types and corresponding operators. The index manager increases fan-out by reducing the entry size for the fast search of large data, and provides an index optimized to the main memory by minimizing cache access failure.

In addition, the data compression manager provides a compression technique suitable for the spatial data characteristics for high network transmission efficiency in data import/export among various systems including telematics applications and middleware, and the transaction manager guarantees data integrity by processing commit and rollback operations, and provides high-performance transaction processing feature in query processing using algorithms optimized to the characteristics of the main memory. The backup/recovery manager provides the function that backups data into the disk using a back-end DBMS to cope with abnormal termination of the system or data loss and recovers the database completely to the state before the failure using the log file provided by the transaction manager.

The data cache manager provides the incremental transmission function that sends important data first and less important data later for efficient data transmission in data import/export, and the import/export manager performs the data exchange function using a standard interface (e.g. ODBC) among various systems including telematics applications and middleware. The memory manager uses the shared memory as a database space for managing large data efficiently in the main memory and provides the memory access synchronization function using the dynamic memory allocation and semaphore based on the first-fit method.

3.2 Data Types and Operators

TMOM supports all spatial data types and operators specified in OpenGIS "Simple Features Specification for SQL" [7]. Also, it supports spatio-temporal data types and spatio-temporal operators by extending spatial data types and operators presented by OpenGIS. Table 2 shows spatio-temporal data types and spatio-temporal operators provided in TMOM.

Table 2. Spatio-temporal Data Types and Spatio-temporal Operators

Geometry Types	Description
ST_Point	Express Point data containing time
ST_LineString	Express LineString data containing time
ST_Polygon	Express Polygon data containing time

Spatiotemporal Relation Operators	Description
ST_Contains(ST_Geometry A, ST_Geometry B)	Return whether Object A contains Object B at each time unit
ST_Disjoint(ST_Geometry A, ST_Geometry B)	Return whether Object A does not meet Object B at each time unit
ST_Equals(ST_Geometry A, ST_Geometry B)	Return whether Object A is the same as Object B at each time unit
ST_Intersects(ST_Geometry A, ST_Geometry B)	Return whether Object A meets Object B at each time unit
ST_Touches(ST_Geometry A, ST_Geometry B)	Return whether the boundary of Object A meets the boundary of Object B at each time unit
ST_Within(ST_Geometry A, ST_Geometry B)	Return whether Object B contains Object A at each time unit
ST_Overlaps(ST_Geometry A, ST_Geometry B)	Return whether Object A overlaps with Object B at each time unit
ST_Crosses(ST_Geometry A, ST_Geometry B)	Return whether Object A intersects Object B at each time unit

SpatioTemporal Analysis Operators	Description
ST_Intersection(ST_Geometry A, ST_Geometry B)	Return the intersection of Object A and Object B at each time unit
ST_Union(ST_Geometry A, ST_Geometry B)	Return the union of Object A and Object B at each time unit
ST_Difference(ST_Geometry A, ST_Geometry B)	Return the difference between Object A and Object B at each time unit
ST_Distance(ST_Geometry A, ST_Geometry B)	Return the distance between Object A and Object B at each time unit

Trajectory Relations Operators	Description
ST_Enter(ST_Geometry A, ST_Geometry B)	Return whether Object A moves from outside to inside of Object B
ST_Passes(ST_Geometry A, ST_Geometry B)	Return whether Object A moves from outside to inside of Object B and then moves to outside
ST_Meets(ST_Geometry A, ST_Geometry B)	Return whether Object A only touches the boundary of Object B
ST_Leaves(ST_Geometry A, ST_Geometry B)	Return whether Object A moves from inside to outside of Object B
ST_Insides(ST_Geometry A, ST_Geometry B)	Return whether Object A exists only inside Object B

TMOM supports temporal data types and temporal operators for processing the temporal data of moving objects. Table 3 shows temporal data types and temporal operators of TMOM.

Table 3. Temporal Data Types and Temporal Operators

Temporal Types	Description
ST_Instant	Express static time
ST_Period	Express a section of time

Temporal Analysis Operators	Description
ST_Intersection(Temporal A, Temporal B)	Return the intersection of Time A and Time B
ST_Union(Temporal A, Temporal B)	Return the union of Time A and Time B
ST_Difference(Temporal A, Temporal B)	Return the difference between Time A and Time B

Temporal Relation Operators	Description
ST_Contains(Temporal A, Temporal B)	Return whether Time B contains Time A
ST_Equals(Temporal A, Temporal B)	Return whether Time A is the same as Time B
ST_Precedes(Temporal A, Temporal B)	Return whether Time A is prior to Time B
ST_Overlaps(Temporal A, Temporal B)	Return whether Time A overlaps with Time B
ST_Meets(Temporal A, Temporal B)	Return whether Time A meets Time B

3.3 Data Compression

To compress spatial and spatio-temporal data, TMOM uses the arithmetic coding technique based on the difference in distance from the standard point. In general, the largest part of spatial and spatio-temporal data in telematics is occupied by

MultiLineString or Polygon rather than by Point or SimpleLineString. MultiLineString or Polygon generally takes a clustered form for a specific region or time. Therefore, if the difference in distance from the standard point of MultiLineString or Polygon is obtained, the data of the actual standard point occupies 8 bytes, but the integer part of difference values appearing afterward decreases gradually. This technique can attain a higher compression rate when moving objects are clustered compactly. Figure 2 shows an example of data compression using the arithmetic coding technique.

Fig. 2. Example of Spatial and Spatio-temporal Data Compression

If +7.8781, the difference between 207068.9921 which is the X coordinate of the first polygon and 207061.114 which is the X coordinate of the first polygon, is expressed in arithmetic coding compression structure, the data can be compressed efficiently. In addition, if not only the differences between the X coordinates of polygons but also -12.2309, the difference between 207068.9921 which is the X coordinate of the first polygon and 207081.223 which is the X coordinate of the second polygon, is compressed, compression performance can be enhanced further by solving the problem in the existing arithmetic coding technique, where the first coordinate of each polygon could not be reduced.

3.4 Indexes

In this paper, we extend existing disk-based indexes, such as spatial index R-tree, spatio-temporal index 3DR-tree and trajectory index CR-tree (Combined R-tree), for the efficient search of spatial and spatio-temporal data. That is, because the existing disk-based indexes are not adequate in the main memory environment, we optimize these indexes fittingly to the main memory for TMOM.

In order to adapt the disk-based indexes to the main memory, CR-tree (Cache-conscious R-tree) is applied with its problems improved. First, to reduce the growing size of MBR, we apply a compression technique that can express the left-bottom point of MBR in relative coordinates (2byte) and the right-top point in the size of MBR (1byte). This technique can reduce the size of data while maintaining the accuracy of MBR appropriately. Figure 3 shows an example that compresses the index of TMOM.

Compressed MBR of object A is (119, 121, 121, 102), and because the size of MBR of object B deviates from the range of 1 byte the MBR can be expressed as (400, 173, 226, 224) through quantization. Second, to reduce the cost of additional update for reconstructing MBR compression, we apply the compression standard point differently for broad distribution and narrow distribution. Broad distribution is a case that the size of the whole MBR is larger than 2 bytes, and narrow distribution less than 2 bytes. In broad distribution, compression is done based on the left-bottom point of extended MBR of the parent node, and in narrow distribution it is done based on the left-bottom point of the whole MBR. If compression is done in this way, it becomes insensitive to changes in MBR and, as a consequence, we can reduce the cost of additional update for reconstructing MBR compression.

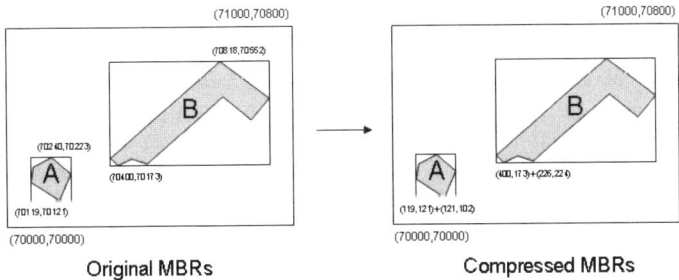

Fig. 3. Example of TMOM Index Compression

3.5 Recovery Technique

TMOM uses the fuzzy-shadow checkpoint technique, which can solve the problems of checkpointing for redundant data and the waste of space in the fuzzy-pingpong checkpoint technique used in most of existing MMDBMS's. That is, using the fuzzy-shadow checkpoint technique that improves the existing fuzzy check point technique based on pingpong update, TMOM records dirty pages in the main memory database into new empty pages in a database file and, if the system fails during checkpointing, it maintains database consistency through recovery using the existing pages. Figure 4 shows the structure of TMOM recovery procedure suggested in this paper.

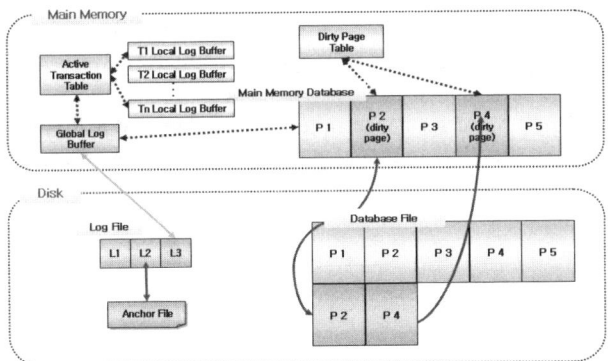

Fig. 4. Structure of Recovery Procedure

In Figure 4, the active transaction table is a table to store information on currently running transactions, and the dirty page table stores information on pages changed after the latest checkpoint. Also, the local log buffer stores the log of each transaction to prevent performance from being lowered by logging competition among transactions, and the global log buffer stores the redo log of partially completed transactions. Lastly, the log file stores the redo log information of completed transactions, the anchor file stores meta information on the log file and the database file when checkpointing, and the database file stores the snapshot image of the main memory database when checkpointing.

4 Performance Evaluation

In this chapter, we evaluated the performance of the index, the data compression, and the recovery technique. The evaluation was made in Pentium IV 2.53GHz CPU, 1GB memory and Redhat 9.0 environment. Our test used data on buildings in Seoul and buildings in a specific area as spatial data, and data generated by the network-based location data generator [2] and GSTD [9] as moving object data.

4.1 Index Performance Evaluation

Index performance evaluation was made by comparing the index size, data insert time and data search time of spatio-temporal index 3DR-tree and TMOM index. Figure 5 shows the comparison of index size and data insert time.

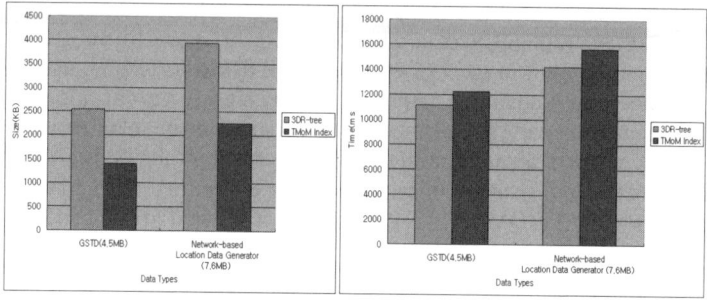

Fig. 5. Index Size and Data Insert Time

As shown in Figure 5, performance in terms of index size is 40~45% higher in the TMOM index than in the 3DR-tree, and performance in terms of insert time is 5~10% lower in the TMOM index than in the 3DR-tree. Figure 6 shows the comparison of data search time.

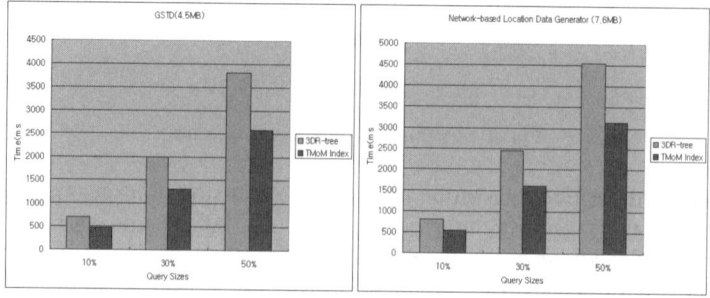

Fig. 6. Data Search Time

As in Figure 6, performance in terms of data search time is 30~35% higher in the TMOM index than in the 3DR-tree for both data from GSTD and data from the network-based location data generator.

4.2 Data Compression Performance Evaluation

To evaluate data compression performance, we compared compression size and compression time between the gzip compression and the arithmetic coding technique used in TMOM. Figure 7 shows the comparison of data compression size and compression time.

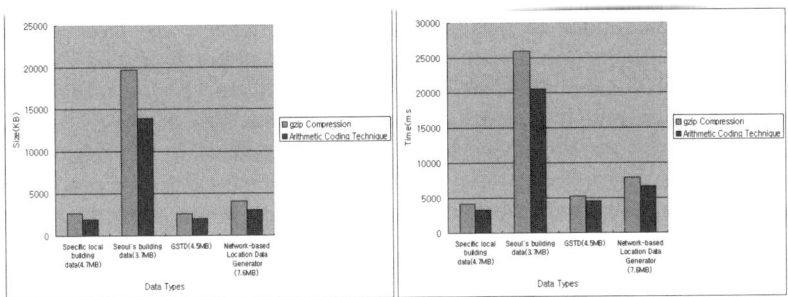

Fig. 7. Data Compression Size and Compression Time

As in Figure 7, performance in terms of compression size is 25~30% higher in the arithmetic coding technique than in the gzip compression for spatial data, and 20~35% higher for moving object data. Also, performance in terms of compression time is 20~25% higher in the arithmetic coding technique than in the gzip compression for spatial data, and 10~15% higher for moving object data.

4.3 Transaction/Recovery Performance Evaluation

To evaluate transaction and recovery performance, we compared transaction processing time, checkpointing time and recovery time between the fuzzy-pingpong checkpoint technique and the fuzzy-shadow checkpoint technique used in TMOM. Figure 8 shows the comparison results of them.

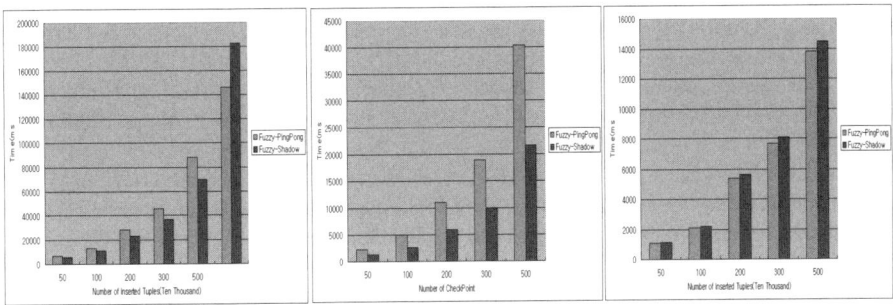

Fig. 8. Transaction Processing Time, Checkpointing Time and Recovery Time

As in Figure 8, performance in terms of transaction processing time is 20~25% higher in fuzzy-shadow checkpoint than in fuzzy-pingpong checkpoint and in terms of checkpointing time 85~90% higher, but in terms of recovery time 1~5% lower.

5 Conclusions and Future Work

This paper designed various spatial, spatio-temporal and temporal data types and corresponding operators for efficient processing of queries in telematics services, and suggested indexing technology optimized to the main memory for the fast search of large spatial and spatio-temporal data. Also, we developed recovery technology that can minimize disk input-output to maximize system performance, and data compression technology for enhancing transmission efficiency in data import/export of TMOM. And lastly, we designed and implemented TMOM that can process and manage moving object data efficiently in telematics services. Furthermore, we evaluated the performance of the developed technologies and proved their superiority. TMOM developed in this paper can be used as a solution for efficient data processing and management in various real-time telematics services. In future research, we need to study and develop duplication techniques for duplicating and maintaining databases to provide the non-stop operation environment when the system terminates unexpectedly.

Acknowledgements

This research was supported by the Ministry of Information and Communication, Korea, under the ITRC support program supervised by the IITA.

References

1. Bohannon, P., Rastogi, R., Seshadri, S., Silberschatz, A., and Sudarshan, S.: Detection and Recovery Techniques for Database Corruption. IEEE Transactions on Knowledge and Data Engineering, (2003) 1120 – 1136.
2. Brinkhoff, T.: A Framework for Generating Network-Based Moving Objects. GeoInformatica, Vol.6, No.2 (2002) 153-180.
3. Guting, R. H., Bohlen, M. H., Erwig, M., Jensen, C. S., Lorentzos, N. A., Schneider, M., and Vazirgiannis, M.: A Foundation for Representing and Querying Moving Objects. ACM Transactions on Database Systems, (2000) 1-42.
4. Kim, K. H., and Cha, S. K., and Kwon, K. J.: Optimizing Multidimensional Index Tree for Main Memory Access. Proc. of the ACM SIGMOD Conference, (2001) 139-150.
5. Mindaugas, P., Simonas, Š., and Christian S.: Indexing the Past, Present, and Anticipated Future Positions of Moving Objects. ACM Transactions on Database Systems, (2006) 255-298.
6. Nam, K. W., Lee, J. H., Lee, S. H., Lee, J. W., and Park, J. H.: Developing a Main Memory Moving Objects DBMS for High-Performance Location-Based Services. Proc. of the 6th Advanced Web Technologies and Applications Conference, (2004) 864-873.
7. OpenGIS Consortium : Inc., Simple Features Specification for SQL, Version 1.1, (1999).
8. Pfoser, D., Jensen, C. S., and Theodoridis, Y.: Novel Approaches in Query Processing for Moving Objects. Proc. of the International Conference on VLDB, (2000) 395-406.
9. Theodoridis, Y. and Nascimento, M.A.: Generating Spatiotemporal Datasets on the WWW. SIGMOD Record, Vol.29, No.3 (2000) 39-43.
10. Yun, J. K., Kim, J. J., Hong, D. S., and Han, K. J.: Development of an Embedded Spatial MMDBMS for Spatial Mobile Devices. Proc. of the 5th International Workshop on Web and Wireless Geographical Information Systems, (2005) 1-10.

Author Index

Printing: Mercedes-Druck, Berlin
Binding: Stein+Lehmann, Berlin

Lecture Notes in Computer Science

For information about Vols. 1–4216

please contact your bookseller or Springer